Nature Conservation

Klaus-Dieter Hupke

Nature Conservation

A Critical Introduction

Springer

Klaus-Dieter Hupke
University of Education Heidelberg
Heidelberg, Germany

ISBN 978-3-662-66158-1 ISBN 978-3-662-66159-8 (eBook)
https://doi.org/10.1007/978-3-662-66159-8

This book is a translation of the original German edition „Naturschutz" by Hupke, Klaus-Dieter, published by Springer-Verlag GmbH, DE in 2020. The translation was done with the help of artificial intelligence (machine translation by the service DeepL.com). A subsequent human revision was done primarily in terms of content, so that the book will read stylistically differently from a conventional translation. Springer Nature works continuously to further the development of tools for the production of books and on the related technologies to support the authors.

This Springer Spektrum imprint is published by the registered company Springer-Verlag GmbH, DE, part of Springer Nature.
The registered company address is: Heidelberger Platz 3, 14197 Berlin, Germany

Preface

After the first edition of 2015 was well received by readers, on the other hand the legal background and the practice of nature conservation are subject to rapid change, the publisher and author have decided to publish a new edition of the work.

Each of the previous chapters has been updated. In addition, a chapter on "Nature Conservation in Times of Anthropogenic Climate Change" was added. A "topical" orientation of the content structure was avoided. For example, the ongoing discussion about insect mortality is taken very seriously and is included in several chapters as an aspect of threatened biodiversity.

The wide distribution of the book in our neighbouring country has been taken into account with a stronger consideration of the situation of nature conservation in Austria, which is similar to German nature conservation in many respects, but still with partly own terminology and contents, and which is even more decentralised than in Germany, i.e. a matter of the Austrian Federal Provinces.

An overall presentation of nature conservation, which would include all relevant aspects and sub-topics, is still not intended. However, feedback to the author has shown that the work is recommended at some (technical) schools and universities as introductory literature on the subject. This aspect has been taken into account in the course of the careful revision; the subtitle has therefore shifted to "A critical introduction". For trainees and students of related subjects, this should suffice. Conservation specialists in the narrower sense, on the other hand, will necessarily need to familiarise themselves with the relevant laws and administrative guidelines during their training. Although these have been taken into account, for the sake of general "readability", they are not the focus of comparable attention, as is the case for the cultural history of nature conservation and its significance for our everyday life.

The book is based on the situation of nature conservation in German-speaking countries, where almost all readers are located. In addition, however, following the first edition, the international area is also taken into account. This is done, on the one hand, by showing the increasing international networking of nature conservation (policy). On the other hand, international regional heavyweights (in terms of area, population and economic power) are presented: the USA, Brazil, Africa, India; but also: the oceans, Antarctica. One could criticise that, in terms of the above criteria, such important areas as Eastern Europe/Russia, China/East Asia or Australia are not taken into account. But the addition would have gone beyond the scope of the work, which should remain manageable and affordable. Here, too, the intention is not unconditional completeness, but the exemplary demonstration of important guidelines and mechanisms of nature conservation.

I would also like to thank the two photographers Wenzel Halla and Stefan Hupke, who contributed photographic material from their travels; and my students, who often had their fingers on the shutter release more quickly during nature encounters on excursions to India, among other places.

Heidelberg, Germany Klaus-Dieter Hupke
Spring 2019

Introduction

It is probably part of the nature of people who are no longer quite young to mourn certain losses that have occurred since their own youth. For me, this is also visible, among other things, in the lack of physical experience of nature by young and growing people. No nature education and no education oriented towards sustainable development have been able to make up for this so far. Children and young people today are driven out of nature to a similar extent as nineteenth century Indian tribes were driven out of their original hunting grounds. There is evidence from anthropology that humans are fundamentally hunters and gatherers and therefore show a tendency to catch interesting creatures and things, collect them and bring them home.

Where this happens today, culturally superimposed, in department stores and on virtual territory, a merely puny substitute for the aforementioned primary experiences has developed.

Of course, when I was young, in the 1960s and 1970s, there were already a few designated nature reserves, and a whole range of animals and plants were not allowed to be picked, caught or disturbed, so they were already protected. However, at that time, virtually no one was interested in what was already prohibited by law.

This has changed significantly in the meantime. Children and adolescents still roam marginal areas, but are now definitely pursued by the legislator or the executive bodies, and in case of doubt the interested public helps. A child who takes home a slow worm or the tadpole of a grass frog inevitably tends to be criminalised ("parents are liable for their children"). In that respect, I feel sorry for today's children. So this book is also written for the young people in my immediate social environment, for Marion and Stefan, for Wenzel, Jonas, David and Jan.

The book attaches importance to a professional (scientific) basic orientation. However, it should also be readable for laymen. Unlike in coleopterology (beetle science) or in atomic physics, there are no experts in nature conservation in the true sense, since every member of society is affected by it and subjective approaches seem important. However, there are certainly people who, for example, know botany or dragonflies a little better than others and can therefore provide important arguments in conservation issues.

In case of doubt, I have sometimes formulated my theses and statements a little more pointedly; not in order to provoke, but in order to stimulate the reader's own reflection.

A complete textbook-like presentation of nature conservation in the German-speaking world was not planned and could not be achieved on the few pages. In this respect, the (now again) specialist will definitely miss some important topics, technical terms and keywords of nature conservation within the presentation. Also the structure of the book is not strictly textbook-like; the individual chapters also do not build on each other. They are largely parallel and can also be read independently of each other, for example in the manner of individual essays.

Otherwise, I am available for personal feedback and also grateful for critical comments.

Heidelberg, Germany Klaus-Dieter Hupke
Spring 2015

Contents

1

What Is Nature for Us?

An acquaintance was very upset some time ago that many neighbors did not like his dog. Today's people would have no more understanding for nature.

Even more than the concrete problems of the friend, I was interested in the association of the dog with **nature**. Of course, the domestic dog *(Canis lupus familiaris)* (Fig. 1.1a) is descended from a wild animal, the wolf *(Canis lupus)* (Fig. 1.1b), and is thus, at least historically-genetically, to be seen in close connection with nature. But are we ourselves, as *Homo sapiens,* not also? After all, the dog, unlike man, has been subjected over millennia to systematic breeding selection, which no one would so generally call nature. And of course the dog also possesses characteristics that point back to its tribal history, e.g. a digestion that only works properly when straying, or a coat that is thicker in winter than in summer. But do not humans also exhibit such naturalistic traits in their physiology and even in their behavior? And if this is so, why is man non-nature?

K.-D. Hupke, *Nature Conservation*,
https://doi.org/10.1007/978-3-662-66159-8_1

Fig. 1.1 (**a**) The wolf as a domestic animal. (Own photo). (**b**) The wolf as a wild animal. (Photo: © Harald Nachtmann)

This example shows that the assignment dog – nature is about something completely different. For us, the dog is nature insofar as it stands for the area not or not always willingly controlled by man. In spite of human education there remains in the dog something unbridled and unpredictable, which in the harmless case expresses itself in an unwanted dog pile (by the way the cause of the above mentioned conflict), in the bad case in the fact that humans are hurt by bites or are even bitten to death. In any case, a dog will always have a tendency to run free, to romp wildly, to prey, etc., which can only be partially or rudimentarily taken away from him. This is the core of the dog's naturalness.

For the **definition of** the **term**, however, it is also appropriate to deal with the absolute opposite of nature in the above understanding as the unregulated and unpredictable. This would then be the regulated and adjusted, the controlled par excellence. We are subject to this controlled in many respects. Just take an average daily routine of a normal citizen. The times of getting up, personal hygiene, meals as well as professional activity with the corresponding breaks are usually strictly regulated. This also applies to leisure activities: on Saturday many people do their bulk shopping, on Tuesday or Thursday they go to a club or self-help group, on Wednesday or Friday they go to the gym, on Sunday afternoon they visit their mother-in-law. For most people, the holidays around Christmas, New Year's Day and Easter in particular are associated with a web of regular routines that are more reminiscent of work than leisure. In addition, there are numerous social pressures and imperatives: shining shoes, brushing teeth (preferably with an interdental brush and dental floss), doing the major cleaning in the apartment, but also not forgetting the bulky waste, sorting the garbage properly, filing the tax return, following procedures and deadlines. Did the kids even pack their school bags properly? And didn't forget their school snack? (I feel so sweaty:) Did I also not forget my underarm deodorant this morning? What about the appointments with the orthopedist and the notary? And what if the orthopedist I'm actually seeing for my hip condition might want to see my feet, too? Didn't one of the socks have holes in it the other day?

This list can be continued almost indefinitely. Their contents are, in a sense, interchangeable. All that matters is that we live in an often unmanageable web of constraints and rituals. Admittedly, these make life easier in some respects. On the one hand, because they are often sensible and useful, at least in social or overall societal terms. On the other hand, however, also because they often run automatically or as conditioned reflexes and thus require little mental or intellectual effort in individual cases. In sum, however, they represent a burden that should not be underestimated. Numerous individual strategies are designed to escape them: We do this, for example, by often staying in bed for "unnecessarily" long periods on Sundays. We escape them in part through "illness", which, especially in the eyes of those around us, gives us a certain dispensation from our daily petty duties and tasks. And we flee from them, again socially regulated, on vacation. Or we flee from them into dreams, both nocturnal and daydreams.

Both in our (virtual) dreams and in our (real) vacations, we repeatedly seek out landscapes that are imagined to be free of people and that obey their own laws. People are foreign bodies in these spaces; if they were to be present in large numbers, they would destroy the character of "nature". These landscapes are at their best and purest when they are exclusively accessible to us alone. The search for **untouched nature** has always formed the main drive for voyages of discovery and exploration, as well as for adventure and self-exploration holidays. The latter are, so to speak, the journeys of discovery for members of today's middle class. Not only is nature visited on journeys, but it is also constructed through these journeys and, in a certain sense, brought forth in the first place.

Since humans need (supposedly) untouched nature as an antithesis and counterbalance to the everyday world that has been transformed and planned by civilization, they tend to regard natural areas as more natural than they actually are. Especially the misinterpretation of foreign habitats and ecosystems such as the North American prairies or the African savannahs as natural (see Chap. 30) bear witness to this. Even the domestic urban forest appears more natural in the light of our needs than it strictly is. And this, of course, also applies to our pet dog, with whom we go into the forest to romp. All of this is nature to us because we need that definition for anything that allows us to shake off social constraints, at least for a moment and largely just as an imagination. The only stupid thing is when we encounter people in this nature who don't like it when the dog leaves its droppings on the footpath, or who object when the dog jumps up on them: people "without a sense of nature".

The fact that we believe ourselves to be under an illusion when, for example, we leave our homes on 1 May or on Ascension Day to go for a walk in nature should therefore be obvious. From our everyday understanding we interpret the forests as well as the agricultural landscape as nature. Occasionally, this view is disturbed by impressions that make the **remoteness of** the respective location clear against all illusion: for example, the screeching of the chainsaw in the forest, the deep wheel tracks made by heavy forestry machines, the wide-open clear-cuts or the tractor spraying agricultural chemicals. The **impression of purest nature is** therefore conveyed by landscapes that correspond to the traditional agricultural landscape before the beginning of industrialization: small-scale, with small plots of land, field hedges and solitary trees, with dry stone walls, ponds and unstraightened streams, with meadows full of flowers in between.

Under these circumstances, however, nature also becomes a synonym for the positive in general, which is contrasted with the unnatural-negative and the alienated (Fig. 1.2 and Table 1.1).

Fig. 1.2 Examples of "non-nature". (**a**) Algae plague on the French coast. (© picture alliance/dpa). (**b**) double-headed turtle "Tom and Jerry" at the Reptile World Anaconda in Halle, 2008. (© picture alliance/dpa – Report)

Table 1.1 Semantics of nature

Defining nature	Contrary term
Area outside social control	The controlled, civilized design
Agrarian relic	The modern, standardized, adjusted, efficiency-oriented
General-positive	The unnatural-negative, the alienated...

Incidentally, children of primary school age are already escaping from the often confining care of their families, building camps or tree houses in forests and creating dams along streams. At least children do this as long as there are still forests or streams in their neighbourhood and they are not driven out of them by nature conservationists, foresters, committed walkers or other "nature protectors".

What Does "Nature" Mean to Us?

For 93% of German citizens, nature simply means "health and recreation". 71% also cited this function of nature as a very important personal reason for the need for nature conservation. When asked about the most important services provided by nature for humans, 26% spontaneously stated "relaxation and recreation"; only "air to breathe" (37%) and "food" (28%) are mentioned more frequently. The recreational effect is thus one of the most present facets of nature in the awareness of the population.

(Bundesamt für Naturschutz 2012d, p. 323; quoted from BMU and BFN 2012 (eds.) Naturbewusstsein 2011).

2

Why Nature Conservation?

Historical Development of the Concept of Nature Conservation

Already in the high medieval Minnesang (around 1200 AD) the enjoyment of nature, especially a spring-like one, played a central role and stood for high quality of life and confidence, the *high muot*. Birdsong and blossoming flowers formed the aesthetic equivalent for a positive inner mood (Fig. 2.1) – and of course, they also already symbolically stood for sexual love or encompassed its ambience. But a demand for **the protection of nature** is not to be found in this or in many other epochs. Not that from today's point of view nature at that time did not need protection. The **High Middle Ages** were a time of population and settlement growth, of degradation and clearing of forests, especially in the low mountain ranges. In the basin areas and lowlands, the forest had already largely disappeared

K.-D. Hupke, *Nature Conservation*,
https://doi.org/10.1007/978-3-662-66159-8_2

Fig. 2.1 The poet Heinrich von Veldeke, who lived in the second half of the twelfth century. (From a somewhat later depiction: Weingartener Liederhandschrift, early fourteenth century). The spatial focus of the depiction, however, is not the poet himself, but elements of nature (tree, birds). It is a spring-like nature: the tree is covered with green leaves; a parent bird on the upper right (yellow) seems to feed a young bird ("colourless" grey). (Recording: Digital Library of the Württemberg State Library Stuttgart)

centuries before – except for the small remnants of small woods that served the predominantly small communities as a winter supply of firewood. Reason enough, therefore, for the demand for nature conservation.

And yet this did not exist, there was not even a rudimentary awareness in this direction. If at all, birds and flowers were seen as forms of nature worthy of protection, which perhaps should not exactly be arbitrarily killed or trampled on. But a principled awareness of the finite nature and vulnerability of nature was not yet present. Perhaps this was also because the processes by which nature was pushed back and destroyed proceeded so slowly that they were not perceived as massive in the course of what was then a mostly short human life. The degraded nature that had been familiar since childhood was perceived as "nature par excellence". A comparative foil, as it is scientifically available today as a construct of potential natural vegetation, for example, was not yet available in a perception that was based more on direct contemplation. Nature was that which existed without any recognisable human intervention. We are no strangers to this view today. Many people speak of the nature they visit on a sunny May Day, for example. This "nature" consists of managed forests, of meadows and pastures that have taken the place of natural forests, or even of grain fields. People in the Middle Ages probably experienced nature in a similarly naive way.

The High Middle Ages were followed by the Late Middle Ages, and then the early modern period with the Renaissance. The return to ancient ideals, however, did not lead to nature conservation in today's sense. Absolutism with its mercantilist economic system also tended to result in **more intensive** use **of nature**, for example in the form of the drainage of moors, or increased arable farming with the omission of fallow land. But such utilitarian thinking is rather opposed to nature conservation.

The decisive **impetus for the protection of nature was** finally brought about by a gradual social upheaval, the significance of which cannot be overestimated. The power of the nobility was gradually broken, first economically. Bourgeois classes, previously at the bottom or in the middle of the social hierarchy, rose and developed a new work ethic. Engineers, often also workers, made groundbreaking inventions, implemented them economically and thus became rich. The stage of **industrialization** was reached. First, as is well known, this process began in Great Britain, but soon it spread to mainland Europe: France, the Benelux countries, Switzerland, Germany. This process of industrialization included not only the actual industry as a production site of goods, but also the entire settlement and **agricultural landscape. The** latter in a double sense: on the one hand, agricultural processes were made more efficient industrially, e.g. with the aid of artificial fertilizers. On the other hand, agriculture was increasingly assigned the role of raw material and food producer for the steadily increasing needs of industrialised society. In this situation of tension, the agricultural landscape changed. It became more efficient, more productive – and more monotonous.

The nobility and the educated bourgeoisie, in some respects the losers of this new development with industrialists and industrial workers as the new social antipodes, reacted insecurely to the loss that occurred. The nobility objectively lost its dominant social position. Less understandable at first, however, is the fact that the now emerging conservationism was supported primarily by representatives of the **educated middle classes. The** latter certainly had its opportunities in the new society: for example, in the increasing number of university chairs and in the emergence of new scientific disciplines. The problem of the educated middle classes was quite different. It was and is interested in the ontological diversity of phenomena, the actual main content of its consciousness: in the diversity of ethnic cultures, languages, local customs and traditions, landscapes, habitats and their animal and plant species.

This (natural) historically determined diversity was drastically reduced with the onset of the industrial age, a process that is still increasing in speed and explosiveness to this day. At first, however, this **loss of** diversity was not so much felt in the distant tropical rainforests or savannahs, and even less in overfished seas, but the educated citizen felt the loss of diversity in his immediate surroundings: for example, during his walks through forests and through the agricultural landscape. Not that these units were still in or near a state of nature towards the end of the eighteenth century, and thus before the onset of industrialization proper on the continent. Quite the contrary: these outmoded agricultural and forest landscapes had been degraded by **overuse** and were often not very productive in an economic sense. Over the centuries, natural forests had developed into extensive hud forests with old oaks, or into anthropogenic heaths or extensively used rough grasslands – in any case, habitats that were almost always surprisingly small in size. This **smallness of scale** resulted in at least two respects: on the one hand with regard to species richness, on the other hand with regard to the optical-visual or aesthetic qualities of landscape. The latter is the view of the landscape painter. The first critics of industrial-era **landscape change** were also very close to art and assessed the changes in the landscape through the eyes of the artist searching for motifs.

In addition to the forestry and agricultural landscape that predominantly surrounded village settlements, the villages themselves and the way of life of the villagers were also subject to rapid change. Local building methods were in retreat, as were local costumes and the multitude of dialects in comparison to the advancing standard language. The cultural life of the villages was also standardized. The university folklore (today: European ethnology) lost its object of research perhaps even more than other disciplines. But also botany and zoology, which, according to the pattern of the time, primarily attempted to record and systematize the diversity of animal and plant species, including corresponding comparative morphological studies, found their objects increasingly rare in the industrially shaped and standardized agricultural landscape. It is no coincidence that (cultural) **monument protection** and nature conservation emerged from around the middle of the nineteenth century largely in parallel in terms of time and also in great intellectual proximity to each other.

The fact that agricultural workers and peasants had not necessarily fared well in the "old days", which were characterized by cultural and natural diversity, was not taken into account. Similarly, the bourgeois critics of landscape change failed to consider that diversity was mostly a consequence of scarcity: of human poverty, which forced the exploitation of even the most marginal opportunities, as well as of the lack of fertile soils or the lack of opportunities to restore the fertility of soils. Nor did the bourgeois observers give much thought to how hard the physical labour in woods and fields might have been; a circumstance that was to be overcome only much later in the course of the industrialisation of agriculture through mechanisation. Their gaze was rather that of the educated layman, the scientist, the artist, the walker and hiker.

But this was also not meant to idyllize the restructuring processes in agriculture dictated by the industrialization processes. For the most part, the farmer had little or nothing to gain. Increasing yields per hectare in the second half of the nineteenth century led to an oversupply of agricultural products and a tendency towards price collapse, which went down in agricultural history as the agricultural crisis. As a rule, the individual farmer benefited from the progress that occurred at best when he changed profession, often by way of generational change, and found a new occupation in industry or in the services. This process, which has also not yet been completed, led to the death of farms and ran parallel to the new forms of labour-saving production, which in turn led to a further homogenisation of agricultural land and thus to new losses in diversity.

From the history of the origin of the term nature conservation, one can learn to understand, on the one hand, that **"untouched nature"** is rarely (in the European context: almost never) meant as the target content. And secondly, that visual and aesthetic bases of perception, individually-subjective or socially-intersubjective, play a major role. In particular, the image of a traditionally transmitted agricultural landscape has become very entrenched in people's minds (Fig. 2.2, contrast Fig. 2.3). Preserving this **traditional agricultural landscape** is still the often unacknowledged goal of nature conservation in Central Europe. It should be preserved even when it has become socially, economically and technically obsolete. The fact that this today requires more work and thus costs more per unit of area than actual (modern) agricultural production is the main problem of nature conservation. In a nutshell, nature conservation tries to preserve the old and is thus, in terms of the history of ideas, something very modern.

Fig. 2.2 Traditionally structured (and species-rich) agricultural landscape using the example of the Randecker Maar (Swabian Alb). The photo dates from the 1970s. (Own recording)

Fig. 2.3 Monotonous agricultural landscape using the example of the Haller Ebene (near Schwäbisch Hall). The photograph dates from about the same time as Fig. 2.2. (Own recording)

A Contemporary's Criticism of the Standardization of the Traditional Agricultural and Forest Landscape

"For the sake of the idea of a convenient straight line, every protruding forest top is shaved, every meadow that extends into the woods is planted full, and even in the interior of the forests no clearing, no forest meadow on which the game could step out, is tolerated any longer. The brooks, which have the bad habit of meandering along in a winding course, must make themselves comfortable to flow straight ahead in ditches ... With the rectangular subdivision of the plots all hedges and individual trees or bushes, which formerly stood on the field marks, then also fall victim to the axe" (Rudorff 1880, quoted after Sieferle 1984, p. 161).

This intellectual modernity contrasts with the integration of the idea of nature conservation into increasingly encrusted conservative world views since the end of the nineteenth century. An "untouched" nature becomes a metaphor in the social and political debate that is "suitable" for very many things; also for idealizing a past epoch or representing "unspoiled folklore". The first comprehensive German nature conservation law was passed by the National Socialists in 1935. The comparatively simple and generous removal of nature conservation areas from intensified agricultural use, which was possible under the new **Reich Nature Conservation Act**, probably took place against the background of the desired land gains in the East.

On the Diversity and Change of Ethical Justifications for Nature Conservation

As was made clear in the previous section, nature conservation is not simply about remnants of a more or less untouched nature, as even interested and educated laypersons often think. Rather, especially in Central Europe, the focus is on old agricultural landscapes, often degraded by centuries of overuse and depleted of nutrients. However, it is precisely these areas that often produce an astonishing biological diversity of animal and plant species, which, in addition to the aesthetic small-scale nature of the landscape impression, is the focus of nature conservation efforts.

Kiemstedt (1967) attempted to quantify aesthetic diversity (Fig. 2.6) in his V-value to assess the importance of landscapes for **recreation.** In particular, a semi-open landscape in which arable fields and grassland are interspersed with small woods or hedgerows, preferably near small bodies of water, has an astonishing constancy in the Western tradition, from the Elysian landscape ideal of the ancient Greeks to the modern English landscape parks (Fig. 2.4). This tradition of the garden is culturally transmitted and may be complemented by an anthropological constant that accounts for the human need for a **semi-open landscape** with a view (to detect prey or enemies early), but also with hiding and retreat opportunities. After all, humans as a species originated in the semi-open African savannahs (cf. also Chap. 30) and in Europe probably first colonized the semi-open forest steppes (warm-temporal) or forest tundras (cold-temporal) in the southeast of the continent.

Fig. 2.4 Diverse half-open artistic landscape of an English Garden using the example of Schwetzingen Palace Park. (Own recording)

The landscape aesthetic diversity reflects the traditional agricultural landscape (cf. Section "Historical Development of the Concept of Nature Conservation" and Fig. 2.2). In many respects, therefore, nature conservation is **cultural landscape conservation** and thus more akin to heritage conservation, which is dominated by historians.

However, one of the crucial objectives of conservation efforts has always been to preserve the diversity of animal and plant species. This diversity is linked to the structural richness of the habitats concerned, which ties them closely to the traditional agricultural landscape. On the other hand, however, it is precisely the nutrient-poor nature of the habitats and ecosystems concerned that gives rise to the diversity worthy of protection, primarily of flora, but also of fauna. This **value of diversity, which is** frequently used today in a striking manner, is perhaps the most original objective of nature conservation and the one that is most easily comprehensible from the outside. Here, at least, it is most likely to do justice to its conceptual-semantic core.

However, the lists of protected species already make it clear that, contrary to ecology, not all species have the same value in nature conservation (cf. Chap. 10). Preference is given, for example, to birds and butterflies, and to orchids (Orchidaceae) among the flowering plants; spiders (Araneae) and inconspicuous cruciferous plants (Brassicaceae), on the other hand, are not given preference. Thus, in individual cases, the value of a species is also measured according to aesthetic standards. This argumentative balancing act is often overcome by elevating attractive species that are easier to find in the field to **target species** and letting the less conspicuous ones swim along as wake species: Where the former are preserved, the latter can often exist unrecognized.

Diversity is discussed in nature conservation at least on two levels. On the one hand, the diversity of species mentioned above is traditionally important. In addition, however, with an estimated 50,000 animal species in Germany (Bundesamt für Naturschutz 2012a; the vast majority of which are insects), it is important to structure this vast number and, if possible, to reduce it operationally. This is achieved by registering **habitats**, **ecosystems** or **biotopes**, in which at least partially identical areas with an expected similar inventory of animal and plant species can be separated out through abstraction. If these can be preserved, a large proportion of the animal and plant species can also be preserved.

A third variant of biological diversity, nowadays usually referred to as **biodiversity,** is increasingly based not only on biotopes and species, but also on the **genetic diversity** below the species. With the ever-improving possibilities of subjecting the genome of animal and plant individuals to comparative analysis, it is becoming apparent that **species** are not as constant and homogeneous as is often assumed. In many cases, they cannot even be precisely identified as such and thus counted. Basically, the assumption of distinct species belongs to a scientifically long-survived world creation model. Only with the existence of distinct species would it have been possible in principle for Noah to include all these species in pairs in his ark. Modern biology has gone far beyond this level of consciousness, at the latest since Charles Darwin. Therefore, the goal is now increasingly to preserve as much of the genetic diversity as possible, even below the species level. Strictly speaking, however, this means that almost the loss of a single individual, at least in theory, contributes to a loss of genetic diversity. The protection of species (one of the main concerns of a conservationist) and the protection of (animal) individuals (concern of the animal conservationist; the conservationist as such rather indifferent) almost meet again here.

A further area of diversity lies in the **landscape-aesthetic diversity** encountered by the naturalist or the recreationist. It is precisely the traditional agricultural landscapes favoured by European nature conservation that exhibit such aesthetic diversity, but also near-natural tropical landscapes in comparison to agricultural monocultures. In the ideal of aesthetic and genetic diversity, the goals of preserving traditional agricultural landscapes merge with the approaches of process conservation and the protection of primary forest sites in the tropics to form a unified objective.

Since the 1980s, the **biodiversity discussion** has shifted strongly to the forests of the humid tropics. Although the number of species and individuals of the large animals of the African savannahs cannot be found there, diversity is often found – even for the trained observer – in the invisible: in the diversity of tree species (which often look the same from below) or in the diversity of insects in the treetops (corresponding research has only been carried out since the 1970s by T. Erwin using insecticides in Panama, by D. Perry using winch systems in Costa Rica, by French researchers using balloon systems in Guyana, by German researchers on cranes in Venezuela and by Malaysian scientists on walkable platforms).

There is hardly a large nature reserve where **research** is not carried out. This is even expressly anchored in the statutes of the national parks. Researchers also enjoy special freedoms in the national parks compared to hikers, for example: They may, functionally appropriate and usually after explicit request, even leave trails and enter the core zone area-wide. In this respect, such protected areas have not only become reserves for threatened animals and plants, but also reserves for researchers who, as "forest-meadow biologists" at our universities, in contrast to the geneticists and molecular biologists who mostly no longer need outdoor research in the field, are also to be regarded as threatened in a certain sense.

Maps of biodiversity are difficult to produce for methodological reasons of data collection and are therefore often shown rather intuitively. For example, it is striking that island archipelagos, especially in the tropics (Indonesia/Philippines, but also the Caribbean), are almost universally attributed a high level of biodiversity, as are the ever-humid to semi-humid subtropics (Mediterranean region, Canary Islands) (Fig. 2.5; Wittig and Niekisch 2014, p. 162). In reality, however, small islands tend to be rather species-poor per unit area compared to, for example, the lowland forests of the Amazon or the western Congo Basin. However, there is a high degree of **local endemism** (exclusive distribution in a very small region); i.e. many species would become extinct if that particular small island no longer existed. Mostly unacknowledged, such local endemism is almost universally highly valued on biodiversity maps. Wittig and Niekisch (2014, p. 162) rightly say, "Biodiversity [is] not only quantity, but also quality." Whereby the latter, which is primarily reflected in local endemism, is particularly difficult to

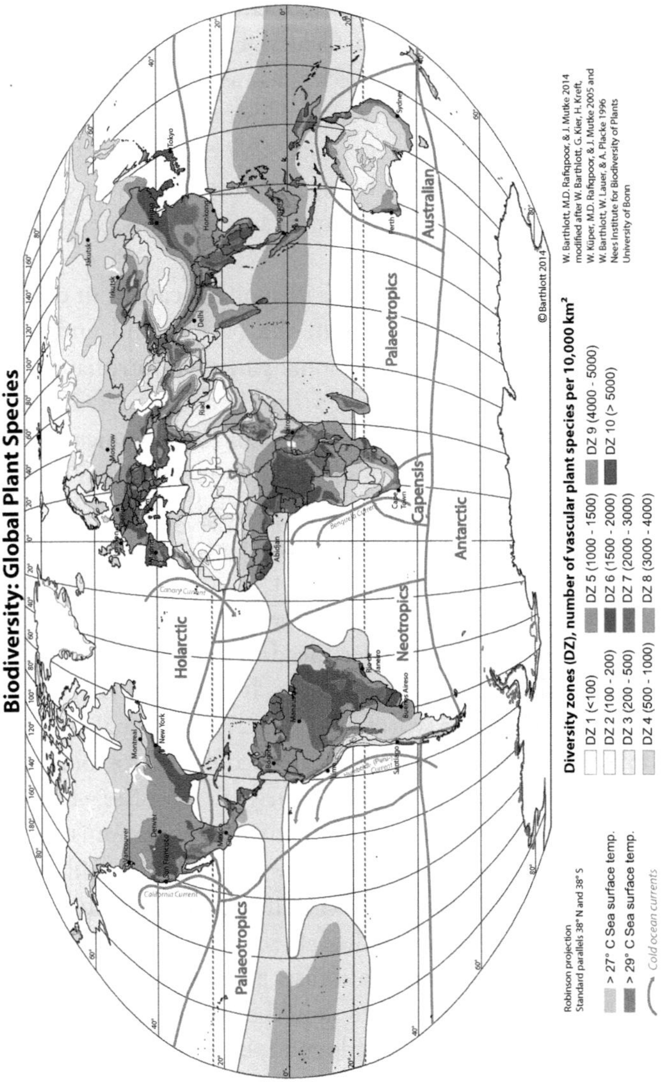

Fig. 2.5 World map of plant species diversity. (© W. Barthlott, University of Bonn; with kind permission)

measure and offset against the former. We have a similar phenomenon of evaluation on Central European ruderal areas, which are often similarly species-rich as the rough grasslands preferred by nature conservation. However, a high percentage of species on ruderal areas consists of widely, often globally distributed ruderal plants, which, since they are not threatened, tend to be rated low in value from a nature conservation perspective (cf. Chap. 13).

In nature conservation, the idea of **education** in dealing with nature is much more in the background and most likely to be integrated in national parks. Unfortunately, the scope of action of the average visitor in a protected area is so limited that the self-initiated forms of nature education often go little beyond the observation of information boards.

In addition, other value bases play a role in nature conservation that are related to mental social escape. In this case, the corresponding habitats in which such needs could be acted out are more likely to be found in "near-natural" forests or "overgrown" river meadows than in the relics of traditional agricultural landscapes. It is also the diversity of demands on nature worthy of protection that makes it possible for nature conservation to encompass not only very different things, but often downright opposites. This depends on which objectives the nature conservationist has in mind in each case (Fig. 2.6). The multifaceted and sometimes contradictory lines of argumentation in nature conservation are summarized in Table 2.1.

In many cases, conservationists also recognise an **intrinsic value of nature**: Nature should be preserved for its own sake. Admittedly, this sounds good, but it is not very coherent in itself. This book repeatedly tries to point out that different forms of nature are possible at exactly the same location, each of which would then have to be granted such an intrinsic value. The dilemma of nature conservation

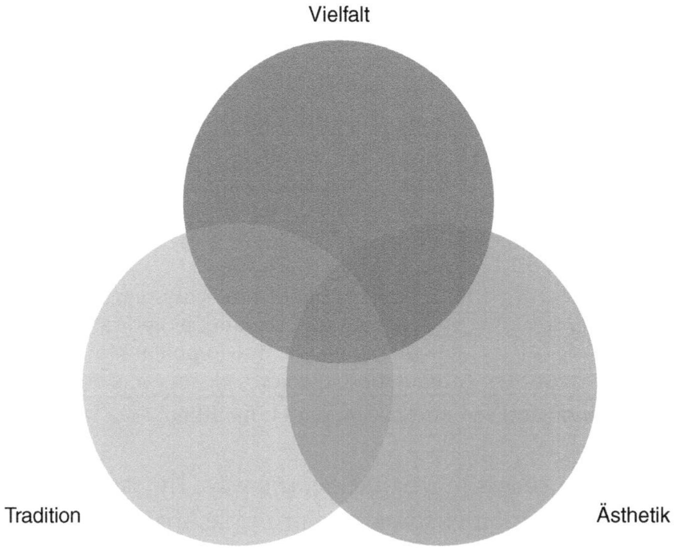

Fig. 2.6 The criteria cloverleaf of nature conservation: diversity, aesthetics, tradition

Table 2.1 Arguments for nature conservation in keywords

Arguments for the nature conservation	Notes on the arguments
Moral and ethical argument	All species have an existence value and a right to exist
Cultural-historical argument	Living cultural assets such as hedges, semi-arid grasslands and their species are just as worthy of preservation as inanimate cultural assets
Aesthetic argument	Many species, biotic communities and landscapes are perceived as beautiful, e.g. orchids, butterflies, species-rich colourful meadows
Health argument	Nature produces pure water, clean air, etc. and is used for recreation
Psychohygienic argument	Visiting nature is good for you: quality of life, sense of home, creation of identity; experience of the original, not man-made

(continued)

Table 2.1 (continued)

Arguments for the nature conservation	Notes on the arguments
Pedagogical argument	Biotopes and species are visual objects and spaces of experience for schoolchildren (and older)
Practical scientific argument	Biotopes and species as objects of investigation for science
Economic argument	Economically important ecosystem services (gene pool for breeding, bioindication, medicinal plants, flower pollination, pest regulation, bionics, tourism value, etc.)

After Gigon and von Rütte 2014, p. 311; modified

cannot be solved in a nature-centric way. The same applies to the protection of species. In principle, many species are in competition with each other; the frequency of animal and plant species as well as their spatial distribution limits are determined at least as much by competition with other species as by abiotic factors. If one assumes the intrinsic value of nature, all these species are of equal value.

We must therefore clarify what kind of nature we want in which location and why we want it (cf. Chap. 7).

Number of the Month

Protecting nature is a human duty – this is how 95% of German citizens see it. When asked about their personal reasons for protecting nature, 67% clearly emphasize the right of future generations to an intact natural environment, 63% are very convinced that animals and plants have a right to exist in their own right, and 50% emphasize responsibility for the global consequences of human actions.

(Bundesamt für Naturschutz 2012b, p. 509).

Objectives of Nature Conservation

- Biodiversity
- Performance and functional capability of the natural balance including the
- Regenerative capacity and sustainable use of natural resources, and
- The diversity, distinctiveness, beauty and recreational value of nature and the countryside

(Bundesnaturschutzgesetz 2009, § 1, in: Deutscher Taschenbuch Verlag 2015, p. 3).

3

Why Nature Conservation Has the Worse Cards Compared to Environmental Protection and Animal Welfare

When television sets became established in German and Austrian homes in the 1950s and 1960s, **animal programmes** were among the most popular types of programming. Initially, it was above all the "television professor" **Bernhard Grzimek** who skilfully managed to combine private interest and conservationist ambitions *(Serengeti darf nicht sterben)* with media presence. Other film authors followed, such as Eugen Schuhmacher with *Die letzten Paradiese.* While Grzimek was very attached to the East African savannah, the edgy Schuhmacher also included other areas of the earth and climatic zones, such as the polar regions. Heinz Sielmann's *expeditions into the animal kingdom* also focused more on native Central European habitats.

Around 1970, a more critical view of the "animal uncle" Bernhard Grzimek took hold among many – starting from the rebellious academic youth – which Loriot wonderfully translated into a sketch. This period saw the emergence of new formats of animal programmes which, for example, did

K.-D. Hupke, *Nature Conservation*,
https://doi.org/10.1007/978-3-662-66159-8_3

not focus on individual species and their way of life, but rather on habitats as such. These included films about coral reefs and about tropical rainforests. This transitional period around 1970 also included the scientifically astonishingly precise programmes of the science journalist Horst Stern, who, among other things, devoted a programme of his own to spiders, which had previously not been considered suitable for television, and was able to present astonishing close-ups using technology that had advanced in the meantime.

Around 1980, however, interest in animal shows began to wane. Programmes with a stronger event character came to the fore, such as sporting events and various "documentaries" and "contests". Early evening series ("soaps") also gained in importance. The internationalisation of German television programming gained in importance, with influences from the USA in particular gaining in significance. These had always relied on strong effects, sometimes neglecting scientific probity (an early example is the Disney production *The Desert Lives*). The decline of the animal film went hand in hand with the rise of private television channels. At the same time, a change in audience structure occurred. The former lovers of animal shows died or moved to old people's homes or nursing homes. The advertising-resistant older audience was no longer of interest to the private stations interested in advertising revenue. In addition, the television culture had changed. Whereas at the beginning of the German television age the family gathered in the living room in the evening to watch television together, today the television is often on "just alongside". Concentrated watching and prolonged observation are becoming less important. This is not a good background for informative nature programmes.

This excursus is intended to illustrate how the **importance of nature conservation is declining**, also in the public eye, and why nature conservation associations are ageing today; a fate, incidentally, that they share with almost all associations with the exception of sports clubs.

The diminishing interest in nature conservation contrasts with the **flourishing of environmental protection in** the 1970s (constitutional status through Section 20a of **the** Basic Law since 1994). Environmental protection is understood as the protection of man's natural environment. Compared to nature conservation, environmental protection is based on a reversal of the model of thinking. While nature conservation aims to protect nature *from humans,* environmental protection aims to protect *the* natural environment *for humans.* This distinction is not a purely academic one, but has quite practical consequences, which Uhrmeister et al. (1998) put into a very illustrative example. According to this, an environmentalist would certainly convert the Rhine Falls of Schaffhausen into a hydroelectric power station. A conservationist, on the other hand, would let it thunder down the rocks as it has always done, completely uselessly. There is no better example of the fundamental opposition between nature conservation and environmental protection.

In the current debate on energy supply, climate and sustainability, this fundamental **conflict between nature conservation and environmental protection** (cf. Chap. 33) becomes apparent in many respects: for example, when tropical forests in Southeast Asia are sacrificed for the cultivation of oil palms in favour of an energy supply with renewable raw materials. In the political-instrumental framework, however, this conflict of interests between nature conservation and environmental protection is regularly ignored. This is evident both in the designation of relevant ministries and in the self-definition of offices such as the State Institute for the Environment, Measurements and Nature Conservation in Baden-Württemberg or the Federal Agency for Nature Conservation with its working levels at the national level. The amalgamation into an overall term "nature conservation and environmental protection" is common practice, and not only in politicians' speeches. There is a danger that the specific interests of nature conservation will be lost.

However, it is not only environmental protection that has gained in importance as a term related to nature conservation in recent decades. **Animal protection** (Fig. 3.1), which like environmental protection is often confused with nature conservation or is not clearly separated from it, has also gained in influence.

Conservationists are concerned with the preservation of animal and plant species as well as the conservation of the habitats concerned. Nature thus interests conservationists at an aggregation level above the animal individual. Animal rights activists, on the other hand, take this animal individual, which is "neglected" by the conservationist, seriously. In a nutshell: animal rights activists demand human

Fig. 3.1 "Calves die for milk" – graffiti on a dirt road. Animal protection, in contrast to nature conservation, has the welfare of the animal individual in mind. (Own recording)

rights for animals. With this fundamental ideal parallelization of humans and animals, animal protection has been quite successful. Perhaps the most successful radical animal protection organisation is PETA (People for Ethical Treatment of Animals). In recent years, German and European animal rights activists have successfully launched legislative initiatives, e.g. to ban the caging of laying hens and the slaughter of dogs and cats (e.g. for consumption or fur production). Since 2002, animal protection has been explicitly enshrined in Article 20a of the Basic Law.

Animal welfare is based on value judgements which, according to Immanuel Kant (1781/1974), are *a priori,* i.e. their reasons are not subject to rational evaluation. However, problems can arise when it comes to the adaptability and practical utility of animal protection ideas. Animal protection, for example, is in contrast to our society's tradition of eating animal foods, especially meat, which is hundreds of thousands of years old. Moreover, the physiology of man and the structure of his dentition alone show that *Homo sapiens* is also a meat eater, although one can of course argue about how high this meat consumption should be and what meat consumption is actually beneficial to man.

The demand for natural living conditions for animals or the call for their "freedom" seems problematic when it concerns domestic animal breeds that would not be viable at all under natural conditions and for which freedom would result in a miserable death. Animal protection deliberately overlooks the fact that both individuals and entire populations of domestic animals have only come into being (and are viable) from the point of view of meat production. To demand for the animal own liberty rights goes thus to a large extent into the void. The situation becomes paradoxical when animal rights activists free caged animals, such as farmed mink, and thus consign them to certain imminent death.

Around 200 years ago, a large proportion of the population were farmers, whereas today in Germany, as in most industrialised countries, this is barely around 2% of the population. The more we move away from a farming society, the more individual attention is paid to the animal, which was seen primarily as a production factor on the farm. **Farm animals** have largely been replaced by **pets** as cuddly animals, which in many ways are more like family members. This has changed the attitude towards animals in wide circles. These influencing factors do not force a turn towards animal welfare, but they do suggest it.

It is interesting that conservation organisations such as the **WWF** (World Wide Fund for Nature, originally World Wildlife Fund) also make use of the animal individual, for example by depicting a polar bear family threatened by climate change on a lonely little ice floe. This is not only a concession to animal welfare, which is not the WWF's core concern at all, but a concession to the viewer, who is assumed to be more moved by animal individuals than a species-based threat scenario. Thus, panda *(Ailuropoda melanoleuca)* (Fig. 3.2), big cats (tiger (*Panthera tigris*), jaguar *(Panthera onca)* and polar bear *(Ursus maritimus)*) have replaced each other as mascots of the world's largest nature conservation organization.

Compared to environmental protection and animal welfare, nature conservation has not yet managed to be enshrined in the Basic Law. Nature conservation can therefore at best be derived indirectly from the constitution.

But there was and is the petition for a referendum in Bavaria, where in spring 2019 well over a million people voted in favour of the protection of bees, forcing the state government and parliament to act. So a great success for nature conservation?

Yes and no. The media-friendly short title of the petition for a referendum "Save the Bees" has deliberately led the

Fig. 3.2 The "cute" panda bears lend themselves as mascots. (© picture alliance/dpa)

majority of signatories to believe that they are standing up for the protection of honey bees. In contrast, the additional text of several pages, which is usually not read intensively, assumes the protection of the several hundred species of wild bees and other insects. The petition for a referendum is thus quite nature conservation-oriented. But due to the difference between the "first impression" and the relevant content, it is also a kind of "deceptive package". Nature conservation uses the public sympathy in favour of the honey bee: pollinator (like wild bees), honey collector and in general "bee-loving", in order to focus "in the small print" on its actual goal, the protection of the many wild bee species. In their sense, honey bees, which have been bred for maximum honey yield for thousands of years, are now rather counterproductive as powerful food competitors of wild bees (Steffan-Dewenter and Tscharntke 2000; Geldmann and Gonzáles-Varo 2018). Moreover, honey bees are not declining at all; on the contrary, beekeeping is increasingly becoming a widespread hobby among urban dwellers.

Supporting the diversity of bee species by promoting honeybees would be like supporting the diversity of large wild mammals by promoting dairy farming.

Animal Protection in the Basic Law

Art 20a.

The State shall also, in its responsibility for future generations, protect the natural foundations of life and animals within the framework of the constitutional order through legislation and, in accordance with law and justice, through executive power and the administration of justice (https://www.gesetze-im-internet.de/bundesrecht/gg/gesamt.pdf ges. 21.4.2017).

Upside Down World in the Forest

A topsy-turvy world, one would think:

Thus, the hunters' guild, which is usually confronted with the accusation of shooting mania, is currently on the defensive when it comes to hunting fever: a nationwide expert opinion, which records increasing game browsing on sensitive tree species such as fir and oak, is countered by the State Hunting Association with a clear refusal to increasingly target the perpetrators of tree damage in the future.

[...]

In contrast, conservationists of all people are sounding the death knell for forest dwellers: because wild boar herds in the state are increasingly devastating home gardens and maize fields, causing car accidents or destroying protected biotopes and thus causing millions of euros in damage, the sows must be put out of business, associations such as NABU have been demanding for years: "Feed less – shoot more" is their request (Die Rundschau für den Kreis Ludwigsburg Jg. 40, Nr. 42 of 16.10.2013).

Author's note: The world in the forest only seems topsy-turvy if you can't distinguish the difference in goals between conservation and animal welfare.

The concerns of nature conservation are also relatively lost in the *sustainability* **discourse**, which has dominated political and media debates since the 1990s. The original intention of sustainable use, which originated with the forestry economist Carlowitz, was not to remove more wood from a (commercial) forest than would grow back in the same period. Admittedly, this is one-sided when measured against the multiple functions of a forest, but it is easy to operationalise, i.e. relatively easy to calculate and monitor.

The problem with the modern approach to sustainability, on the other hand, is that this operationalizability does not seem to exist. Generally formulated ecological goals are interwoven with economic, social and cultural objectives, so that ultimately goals and paths appear vague and sustainability has very much taken on the character of a positive "catch-all".

If representatives of this approach nevertheless formulate concrete ideas, these tend to serve the environmental protection goals already outlined rather than those of nature conservation – with the corresponding problems.

4

Nature Conservation: On Which Areas?

Nature conservation requires areas. The question arises whether a (necessarily less intensive) protection on a large area or an exclusion of smaller areas as intensively protected "nature reserves" is to be aimed at. It is quite nice, if some small details are protected, however, this nature monument work has no meaning for the general public. Tittle-tattle is nature conservation as we have it. Nature destruction, on the other hand, cannot be denied an ingenious generosity. Nature destruction works "en gros", nature conservation "en detail"—(Hermann Löns 1911; quoted after Schönichen 1954, p. 279).

What Hermann Löns criticised here more than 100 years ago was the **small size of** the areas protected by "nature conservation", measured against the extent of the other agricultural and forest landscapes. In the course of modernization and mechanization of their cultivation, these vastly predominant areas are becoming increasingly standardized and thus correspond less and less to ideas of nature conservation.

K.-D. Hupke, *Nature Conservation*,
https://doi.org/10.1007/978-3-662-66159-8_4

Compared to the first beginnings of nature conservation in the nineteenth century, it has been possible in the meantime to continuously expand the proportion of protected areas, but even today it still amounts to a few percent of the total area at most (nature reserves and national parks as strictly protected areas account for a total of around 4% of the state areas in Bavaria and Baden-Württemberg, for example). However, since at the same time the intensification and standardisation of the agricultural landscape in particular have increased even more drastically, the discussion about the land **preference of** nature conservation and land use remains largely intact.

In addition, biological genetics has shown that, at least in the long term, **populations** can only survive if a certain number of individuals are present, because otherwise gene loss sets in. Modern conservation approaches therefore attempt to link protected areas with each other. Ideally, this is done through linear or band-like structures. Stream courses and their accompanying riparian vegetation, often strips of woodland, are ideal for this purpose. However, empirical evidence of the benefits of such **linkages** in terms of stabilising species diversity is scarce. Moreover, the diversity of individual biotopes and habitats must also be taken into account. As a linear structure, a protected floodplain is certainly well suited to linking stream species. But does this also apply to species of rough grasslands, for example?

Another design of nature conservation planning is the **stepping stone concept:** related or similar habitats should be located in such close proximity to each other that the respective populations can exchange without problems. The concept is obvious in its logic, but would require extremely extensive research into which of the tens of thousands of animal and plant species native to Central Europe are now able to bridge which distance. It can be assumed that species capable of flight, such as birds and most insects, are able to do this over at least several kilometres, whereas amphibians and reptiles are not. The characteristics of the areas between the stepping stones must also be taken into account. There are landscape types that can be walked through, even if they are unsuitable as permanent habitats for most species. Highways, however, are certainly not one of them. On the whole, transport is well connected for obvious reasons, whereas nature-protected areas are mostly very isolated.

In many cases, there has been a complete departure from a policy of designating nature conservation priority areas and calls for nature conservation over the whole area. In practice, this will only be possible if considerable compromises are made with regard to agricultural and forestry economic efficiency. Basically, this can only be achieved by extensifying agricultural production.

The conditions for such **extensification of** agricultural and forestry production were favourable in Europe for decades. Agriculture produced surpluses in the sense of "cereal mountains", "butter mountains" and "milk lakes". The agricultural policy of the 1980s and 1990s was characterised by a move away from the promotion of production towards the promotion of individual farming livelihoods. This could be well combined with nature conservation goals.

For some years now, however, we have been seeing the opposite trend. The expected global warming has led to increased **promotion of renewable energy materials** in almost all industrial societies, in Germany accelerated by the amendment of the Renewable Energies Act of 2009 (cf. Chap. 33). An ever greater proportion of agricultural production is being used for energy, leading to a shortage of agricultural goods and a trend towards rising prices, also worldwide. Apart from short-term fluctuations in harvests, agricultural prices, which have traditionally been stable in the long term, are increasingly being linked to crude oil prices in market terms, with strong price increases expected in the future. Agricultural land is becoming increasingly scarce and expensive. The likewise increasing use of wood reserves in forests for heating systems points in a comparable direction.

Nature conservation must therefore compete with **land use** for land and is thus becoming increasingly expensive. Once again, the competitive situation between environmental protection and nature conservation, which is little perceived by the public, becomes apparent here. The demand on nature conservation, which is less popular anyway, is obvious, but to put its goals on the back burner somewhat in order to protect the "more important" world climate.

As a consequence of the above-mentioned trends, the agricultural landscape is developing towards spatial segregation: a small percentage of the area is being taken out of intensive use and managed primarily under the interests of nature conservation. The proportion of land that is at least twenty times larger is being subjected to increasing intensification pressure, with high use of agrochemicals and ever decreasing biological diversity (see Chaps. 32 and 33).

In principle, **agricultural use** is not excluded in **nature conservation areas**; according to the Federal Nature Conservation Act (and according to factual logic), it may only not run counter to the purposes of nature conservation. These purposes result from the protection statutes, which must be specifically defined for each nature conservation area. In a comparison of different nature reserves, these can even be opposite objectives: for example, in the case of a complete withdrawal of man, a succession into the "state of nature", in the other case, careful maintenance in order to preserve the intended state. Often this **maintenance,** which imitates traditional agricultural use techniques, is even indispensable for the preservation of the condition worthy of protection. A large part of the light-open rough grasslands, heaths and fens must be kept artificially free of forest by constant mowing.

However, such a yield-intensive farming method is no longer competitive from the point of view of efficient (= low-cost) agricultural production. Today, high yields are produced per unit of land, above all with enormous expenditure on fertilisers. However, this fertilisation – and thus any intensification of agricultural use – favours a few particularly fast-growing plant species, which then displace the great diversity of other species. Nature conservation and high agricultural yields (the latter both per area and in the overall balance of farms) are thus mutually exclusive. While traditional agriculture in a certain sense brought about this high biodiversity in the first place, modern agriculture has become the antagonist of nature conservation.

To a certain extent, a link between the designation of priority nature conservation areas and overarching overall land protection is provided by the field **margin strip programmes** (cf. also Chap. 25), which have been adopted by most state governments in Germany since the 1990s. Farmers receive money for removing marginal strips from intensive agricultural use, especially along field paths. Although the areas concerned are also clearly demarcated and defined here, an effect is nevertheless tangible that is more strongly related to the entire agricultural area than the designation of natural monuments and nature reserves. In addition, the linear structure of these biotopes facilitates their spatial connectivity, which, among other things, favours the migration of species.

Natural monuments in particular, as the smallest units of protected areas, often include only individual species, as in the case of the protection of individual old trees. Here, provided that the tree species is very common and not threatened in any way, the aestheticizing character of many conservationist approaches becomes apparent. Even if a tree naturally requires a certain area for its mere physical existence, this is not a matter of protecting an area, but rather of protecting a plant individual that also provides a small habitat for other living creatures, especially birds and insects.

Nature Conservation Through Contract Farming and Extensification of Land, Using the Example of North Rhine-Westphalia

As early as the late 1970s, there were efforts in North Rhine-Westphalia to couple agricultural use and biotope maintenance. In a three-year pilot project starting in 1985, the Geobotany and Nature Conservation Department was commissioned by the North Rhine-Westphalian Ministry of the Environment to test whether this idea could be put into practice – and if so, at what cost. Contracts were then concluded with around 40 farmers for the renaturation and nature-conserving use of meadows, pastures, rough grasslands and heaths.

At that time, the contract nature conservation areas in the Eifel comprised just 200 ha. Today there are about 3600 ha. In the meantime, there are more than 500 farmers in the districts of Euskirchen, Düren and Aachen who successfully operate contractual nature conservation, some of them for several decades. In the whole of NRW, about 5200 farmers participate in contractual nature conservation on more than 25,000 ha.

Spectacular Success

From the beginning, the effects of the measures were scientifically monitored with funding from the North Rhine-Westphalian Ministry of the Environment, primarily by the Bonn Faculty of Agriculture. For the research project now completed, the researchers evaluated around 150 Bonn examination, diploma and doctoral theses from the last 30 years. They compared this historical data with current surveys of more than 100 areas in the Eifel. "Such a favourable data basis is probably unique in Germany", says [the supervising scientist; d. Vf.] Schumacher not without pride.

The results exceed all expectations: Not only could the creeping loss of plant species be halted – in the Eifel, the previous trend has even been reversed: most Red List species show strong, sometimes even exponential increases in their populations. The species diversity on the meadows and pastures has also stabilised and in some cases even increased by a factor of two to three. The situation is similar in parts of the Siegerland and Hochsauerland regions, for example, where contractual nature conservation has also been practiced for more than 10 years. Professor Schumacher: "These regions are probably among the few cultural landscapes in Germany where a turnaround has been achieved." The success story was made possible, among other things, by the willingness of the state, districts, municipalities and the NRW Foundation to renaturalise their areas and lease them at favourable prices.

Equally important, however, was the financial support for contractual nature conservation from the state of NRW and the European Union. "However, the principle of 'nature conservation through use' can only be sustained if what grows there is also utilised on the farm," emphasises Schumacher. "This integration of nature conservation into the agricultural business is an essential part of the concept" (Verband Biologie, Biowissenschaften und Biomedizin, VBIO e. V., https://www.vbio.de/?news_id=4089Viele; ges. 24.2.2015).

5

Extreme Locations: Shunned by Industry, Preferred by Nature Conservation?

As Chap. 2 made clear, it is primarily aesthetic landscape reasons that lead to the protection of selected landscape elements. Insofar as these nature conservation areas or (in the case of more selective extension) natural monuments are and were based on pre-societal natural structures, they are disproportionately special sites such as rocks, waterfalls, caves and calcareous tufa grottos, boulder dumps/rock seas (Fig. 5.1), dunes and other landscape elements which are therefore rather atypical in the area. Most of them grow on shallow soils, which are neither suitable sites for agriculture nor for intensive forestry. Extensive flat areas are also generally preferred for large factories and cultural and recreational facilities such as schools, baths and sports halls.

The protection of rocks and other **extreme sites** is often easy because, apart from possible use by quarrying companies, there is often no question of economic use, i.e. the corresponding areas are already cheap in market terms. An important prerequisite for nature conservation, not to cost

K.-D. Hupke, *Nature Conservation*,
https://doi.org/10.1007/978-3-662-66159-8_5

Fig. 5.1 Boulder field extreme site is one of the few naturally largely forest-free areas in Central Europe and, with its characteristic small fauna and flora (including other organisms such as lichens), is often designated as a nature reserve or natural monument. (Small Odenwald near Heidelberg, own photo)

much, is thus already fulfilled. In addition, these sites are often rich in specialists in flora and fauna, which keep the corresponding biodiversity high. And finally, such sites are usually less deformed by man than economically more suitable sites with deep soils. The economically ideal case of a natural monument is a cave: Since it is purely underground, it by no means precludes land use on the surface of the earth on the floor above.

It is therefore not without reason that the largest protected areas in Germany have become the extensive **Wadden Sea areas** on the North Sea. Without large and therefore mostly uneconomical cultivation efforts, these areas, which are cyclically occupied by the sea, can hardly be used intensively. And yet these tidal flats are rich in species and biomass. They are therefore ideal for designation as nature reserves. One only has to be careful to leave a few loopholes open for the marginal interests in terms of area: for example, for crab fishing or for tourist mudflat walks. Otherwise, nature conservation in the Wadden Sea is more or less free of charge – and improves the area-related German nature conservation statistics enormously! For politicians, such performance figures often form the yardstick for nature conservation action and can serve well alongside the inflation rate, economic growth and unemployment rate as a numerical argumentation support for the political performance balance.

Naturally, areas that have been heavily anthropogenically altered are also particularly suitable as nature conservation areas, provided they cannot be used commercially in accordance with their character. This was the case, for example, for the Dümmer in Lower Saxony, which served as a flood reservoir after its embankment during the National Socialist era. It is only in recent times that a conflict of interests has arisen here due to an increase in the use of the water areas for tourism.

However, the sewage fields of large urban sewage treatment plants, which often provided an amphibious habitat for waterfowl and marsh birds as well as resting places for migrants, could hardly be used for purposes other than nature conservation. Due to the progress of wastewater technology, most of the sewage fields have now been built over (as in Freiburg i. Br.) or have at least largely lost their original function of wastewater treatment (as in Munich).

6

Confusing Diversity: Area Categories of Nature and Landscape Protection—Nature Reserves, National Parks, Natural Monuments, Landscape Conservation Areas, Nature Parks

For people (even educated ones) who are not regularly involved in nature conservation as a hobby or profession, a national park and a natural park are quite easy to confuse. The common feature of both categories is that they are very extensive in terms of area: several dozen to several 100 km^2. This means that they cover an area ranging from the size of an average municipality to the size of a county.

However, this also largely exhausts the similarities between a national park and a nature park. While a **national park is** subject to the strictest nature conservation conditions in its core area and may not be entered without a special permit, a **nature park** is not much more than a kind of deceptive packaging that conveys a tourist-effective label rather than a specific protection status. However, every German nature park probably contains areas with a stronger protection status in some areas, especially nature reserves. But a nature park also includes areas that are not subject to any protection. This special procedure has made

K.-D. Hupke, *Nature Conservation*,
https://doi.org/10.1007/978-3-662-66159-8_6

it possible for large parts of our low mountain ranges to become nature parks without the initially sceptical nature park communities being significantly restricted in their planning sovereignty.

A **nature reserve,** on the other hand, is an area where the protection of nature has priority. Agricultural or forestry uses are by no means excluded. In fact, they are often a prerequisite for maintaining the status quo, which is considered worthy of protection. In order to ensure optimal protection, each nature reserve must have its own nature conservation ordinance, which also regulates the possibilities, type and intensity of any use. Nature reserves are therefore individuals that must not be treated in the same way according to the watering can principle. Where in one case the overgrown biomass of a rough grassland has to be mowed and removed every year, in the other case of a forest reserve the area may remain unaffected for decades. Which management measures are adopted or not is determined by the desired final status of nature and the interpretations of how this can best be achieved or maintained.

On the one hand, nature conservation areas represent a separate type of protected area within the meaning of the German Federal Nature Conservation Act. On the other hand, however, the strictly protected **core zones of** a national park or a biosphere reserve are also counted as nature conservation areas. In this respect, nature conservation areas form a super-category that applies to all strictly protected areas: i.e. those in which nature conservation has priority.

Because nature reserves very much restrict use, their designation is usually only possible if they are public property (or those of relevant foundations or organisations). Anything else would mean the expropriation of the previous owners and users.

Actually, a **natural monument** is also such a nature reserve, albeit a small **one.** According to the German Federal Nature Conservation Act, an upper limit of 5 ha is set for a natural monument. If it is larger, it is a nature reserve in the narrow sense. However, natural monuments do not have to be defined in terms of area at all. A single tree or a geological outcrop, for example, can also be protected as objects, in this case without reference to area.

For more than 100 years, natural monuments have been subjected to widespread criticism as “Pritzelkram” (Hermann Löns) by conservationists who want to protect nature over a large area, if not over the entire area. The first protected objects such as the Drachenfels in the Siebengebirge near Bonn were initially small in size. The first nature conservation office in Prussia was the Staatliche Stelle für Naturdenkmalpflege (founded in 1906). Nevertheless, it does not seem unreasonable to place small objects of nature that appear worthy of preservation under special protection as such. If these monuments are particularly numerous and interrupt and complement the intensively used cultural landscape, they can certainly make a contribution to both aesthetic and species diversity. Massive designation of natural monuments should be possible, since the low land use in individual cases makes the inhibition threshold for designating a new natural monument appear low. Intensive enforcement with protected small elements such as individual trees, hedges, dry stone walls, etc. also enhances the landscape section in question in terms of its broad area.

A **landscape protection area** is an area where only the rough character of the zoning may not be changed. This means that a forest may not be cut down here or agricultural land may not be afforested. Also settlements may not be extended into a landscape protection area. However, a landscape conservation area does not protect against radical conversion or intensification of a given land use category. A semi-natural forest may be converted into a Douglas fir forest. Grassland may be turned into a horizontal maize field. This is possible without any problems in a landscape conservation area, unless other regulations speak against it. Accordingly, the possibilities of use of agriculture and forestry are only slightly restricted and the protection of nature is of secondary importance. Even the landscape protection contained in the term is only guaranteed in a very broad understanding of the same.

In principle, the following applies: Precisely because the landscape conservation area imposes only a few conditions on the land owners and users and largely guarantees the previous type of use, including intensification of use, it can also be designated over a wide area. Large areas of woodland and agricultural land, especially in our regions close to large cities, are now landscape protection areas, whereby these primarily act as a brake on settlement expansion. However, the following also applies just as fundamentally: the more intensive and strict the protection requirements of a particular protection category are, the smaller the designated areas will generally be.

Because landowners can continue to use their land, landscape conservation areas have largely remained in private ownership.

Again strictly protected, but on a small scale, are total reserves of the Forestry Commission (Fig. 6.1), which are called **Bannwälder** in Baden-Württemberg, but are also common in other federal states under several other names; in Austria, where "Bannwälder" stand for avalanche protection forests on mountain slopes, under the name **Naturwald** reservat. Here man has completely withdrawn from using and influencing the forest vegetation. The aim is primarily

Fig. 6.1 Bannwald in the middle Neckar region north of Stuttgart. The fact that the forest is no longer used can be seen from the remaining strong copper beech (*Fagus sylvatica*) on the left; also from the fact that the dying lime (*Tilia* spec.) on the right, which leans to the side, is not removed, as well as from the deadwood in the foreground and middle ground on the right. Also the strong young growth, especially of sycamore maple (*Acer pseudoplatanus*), remains standing and hinders the continuity of the forest. (Own recording)

a scientific one: to investigate which type of forest is suited to the soil and climate at a particular site and will prevail in the long term. However, given the longevity of trees, one will have to wait a few centuries before (perhaps) the expected scientific gain in knowledge is achieved. "Perhaps" because vegetation research has increasingly moved away from a fixed **climax vegetation** as the desired final state under certain environmental conditions (in the German-speaking world mainly modified as: potential natural vegetation). Today, especially in species-rich forest ecosystems, the expected developments are considered to be, at least to some extent, open and unpredictable. This is due simply to the introduction of species that are originally alien to the site, but which occasionally develop a high competitive power.

Bannwald, Natural Forest, Wilderness Area ...

From this point on, the term natural forest is used for smaller or larger areas that have been completely removed from forestry (or other) use. The term Bannwald, which is commonly used in Baden-Württemberg, is striking, but has the disadvantage that in the Alps this term is used to describe avalanche protection forests above villages, which are open to economic use. The term natural forest is used as an official designation in North Rhine-Westphalia and Saxony, among other places, but also in Austria.

Since the aim of protected forests/natural forest reserves is to show the subtle differences in location, they are often scattered over a large number of sites, but are therefore only possible over small areas in order to avoid having to remove too large a proportion of the forest from the forestry use concept. This small area alone makes a rapid natural change to a primeval forest appear questionable, since an ongoing influence from neighbouring areas is to be expected – for example through game browsing in the mostly unnaturally high populations of cloven-hoofed game in Central European forests due to the extensive extinction of the "large predators" brown bear, wolf and lynx.

As Bannwälder are strictly protected, their area is also included in the super-category of nature reserves.

Since 2005, the large protected areas **national park, biosphere reserve** and **nature park have** been combined to form the **National Natural Landscapes**; organized by the Brussels-based *Europarc Federation* as an umbrella organization. This term is not mentioned in the current version of the German Bundesnaturschutzgesetz (Deutscher Taschenbuch Verlag 2015). It is to be seen as a rather unofficial label that serves the better self-presentation and marketing of the major categories of nature and landscape conservation, which otherwise have few common features.

The Most Important Area Categories of Nature and Landscape Protection in Germany (Terms and Criteria in Austria are Comparable)

Bannwald/Naturwald (various names depending on the federal state): Completely unmanaged trial areas of the state forestry administration, which are to develop into "primeval forest" again in the long term; mostly small areas (a few hectares).

Biosphere Reserve: Similar to National Park in size and zoning, but does not primarily serve nature conservation, but rather the development of sustainable economic practices, assigned to the *Man and Biosphere Programme of* UNESCO.

Landscape protection area: Only the basic categories of land use (forest/agriculture/settlement) may not be expanded or merged. Otherwise not subject to strict conservation criteria. Largely privately owned.

National Park: Large area, usually spanning several municipalities; divided into core and peripheral zones; only the former strictly protected, the latter also used for tourism.

Natural monument: Like **nature** reserve, but small area (according to BNSchG <5 ha).

Nature park: Large area, usually also includes nature reserves and landscape protection areas, otherwise without special protection status, overall rather label for better tourist marketing.

Nature reserve (in the narrow sense): Mostly in public ownership, nature conservation has priority over agricultural and forestry use.

National Park in the Northern Black Forest: Pros and Cons

Pro National Park

A piece of wilderness, 100 km^2 in size and thus comprising just 0.7% of Baden-Württemberg's forest area, is stirring up tempers in the northern Black Forest. The area of the planned national park lies entirely in the state forest, and a large part is already protected. These islands of nature conservation are to be linked and three quarters of the area left to the mercy of flora and fauna after 30 years of development. Skiing, cycling and hiking are permitted on marked trails; people are expressly welcome in the national park as sportsmen, bon vivants, observers and learners.

Apart from Rhineland-Palatinate, Baden-Württemberg is the only federal state that does not yet have such a protected area. The green-red state government wants to change that. The independent expert opinion, which was put out to tender throughout Europe, provides the factual arguments for this. Its conclusion: a national park in the northern Black Forest would offer the region many opportunities as well as added value for nature, tourism and the economy. The risks – for example from bark beetles – are controllable. The government sees this expertise as confirmation of its plans (Andrea Koch-Widmann, *Stuttgarter Zeitung*, 9.4.2013).

Contra National Park

In a rationally weighed overall assessment, it becomes abundantly clear that the expected adverse consequences of a possible national park in the northern Black Forest far outweigh the claimed, but overall either minor or refutable advantages; in particular, there are no convincing, scientifically provable arguments for a large-scale protected area such as the so-called national parks, either from the point of view of nature conservation or from the point of view of tourism. While the old-growth and deadwood concept developed by the Baden-Württemberg State Forestry Service for the entire Black Forest can be expected not only to safeguard but even to improve biodiversity while at the same time optimally balancing the interests of all forest stakeholders, market-driven hotel offers and an attractive tourist infrastructure are decisive for tourism success, not, however, the designation of a large-scale protected area that is only apparently unaffected by human activity.

The possible establishment of a national park would thus be an exclusively political decision to satisfy the interests of numerous nature conservation associations.

(Wolfgang Tzschupke 2013, National Park "Nordschwarzwald" – Argumente und Gedanken. Gutachten zu den möglichen Auswirkungen eines Nationalparks im Nordschwarzwald; website: https://www.unser-nordschwarzwald.de/wp-content/uploads/2013/03/NP_NSchwarzwald_Gutachten_2013.02_c.pdf; ges. 6.8.2014).

7

Which Nature Do We Want to Protect and How?

Let us go back in history more than 7000 years, to a time before the first clearings took place in Central Europe. Almost the entire area of present-day Germany and Austria was covered by forest at that time, predominantly mixed deciduous forest, depending on the local bedrock, exposition, relief, climate and soil in a specific mix of tree species in each case.

The precursors of today's cereals were cultivated on the land that was later cleared. Since there was no sufficient and systematic fertilization, the land had to be left fallow again and again for years. In addition, there were the extensively grazed areas at the edges of the village farmland. Overall, man created a **mosaic** of open areas and areas still covered with forest. In the open areas, species migrated that had previously not found a place in Central Europe, at least over large areas: light-loving species from the high altitudes of the Alps or from extreme rocky locations, species from the steppes of South-Eastern Europe or from the Mediterranean region. Since the actual forest species mostly did not die

K.-D. Hupke, *Nature Conservation*,
https://doi.org/10.1007/978-3-662-66159-8_7

out, this **immigration** brought an increase in the number of species and communities (and thus a higher beta diversity). Today, this can give the impression that these ancient migrant plants (**archaeophytes**) and ancient migrant animals (**archaeozoans**) have always been part of our landscape.

But this development, which had reached a stable equilibrium over thousands of years, came under threat from the nineteenth century onwards, and increasingly so from the twentieth century onwards. The more intensive land use techniques mostly endangered these old immigrants rather than the actual species of the forest. Herbicides and insecticides tended to eliminate from the agricultural landscape everything that did not serve its production purpose. Small forms of the established cultivated landscape such as pools, field hedges and solitary trees tended to be perceived as obstacles in the age of machine work and were usually removed. Increased inputs of fertilizers favored the growth of a few particularly vigorous species at the expense of the many others that became rarer or extinct.

The consequence of this process of change is a rapid **decline in species numbers in** the agricultural landscape. This is perceived by nature lovers (Chap. 2). This results in the paradox that the focus of nature conservation today is often more on (pre-)historically immigrant species than on the species that actually occur naturally.

This situation clearly shows a **target dilemma of nature conservation**: What should be protected at all? Figure 7.1 uses the example of sand dunes in south-west Germany to illustrate the diverse potential of nature. Does nature conservation mean protecting nature that is as "untouched" as possible or, if necessary, restoring it? In this case, humans would have to withdraw as far as possible from the protected areas. This would be "nature conservation by omission". Another objective of nature conservation is interested in diverse and in turn species-rich biotic communities. The

Fig. 7.1 (**a**) Tall mixed forest of pines (*Pinus silvatica*) and oaks. (**b**) Open sandy grassland of mosses, lichens (Lichenes) and herbs, both at a comparable site and at the same time of year (end of April). (Own photographs)

sum of the existing animal and plant species is the goal here. In contrast, aspects of the naturalness of a biocoenosis take a back seat. The focus here is less on original nature and more on an agricultural landscape that has been handed down and is itself endangered.

Such relicts of a **traditional agricultural landscape** have been able to survive above all in unfavourable and peripheral locations, which have always been used more extensively than favourable areas. Such an unfavourable **location** – from the point of view of potential agricultural yield – can arise within the framework of different natural units.

In the rainy north-west of Germany these are mainly the sandy soils washed out by lime and often also by clay constituents, mostly remnants of the older ice ages. The clearing of the forest, especially for the supply of firewood to the salt works of Lüneburg, has accelerated the process of degradation and loss of nutrient salts. What remained was a botanically rather monotonous landscape, the **heath,** which was often dominated by heather (also known as broom heath, *Calluna vulgaris*). Peasants of the Middle Ages and the early to mid modern period may well have found this type of landscape repulsively ugly. Heath semantically means "barren", lacking productivity. This was also due to a hard so-called hardpan (caused by sedimentation of the surface washed minerals) of the podsol soils formed under these conditions at a depth of a few decimetres. This is partially waterlogging and unfavourable for agricultural use. It was not until the second half of the nineteenth century that steam ploughs were used to break up the stony horizon and, with subsequent fertilisation and liming, the heath was also made productive for arable farming. However, the disappearance of the heath also marked the beginning of its romantic exaltation, which ultimately led to the early protection of the last remnants of heath around the Wilseder Berg as the Lüneburg Heath Nature Reserve. Today, there is a mass influx of visitors, especially in the months of August and September, when the heath is in bloom.

From a botanical point of view, the Atlantic heath is quite poor in species. The naturally nutrient-poor sands of the past ice ages, on which the northern German heath landscape has developed preferentially, lead to an extreme **lack of nutrients**, especially calcium, potassium, magnesium, nitrogen and phosphorus, which precludes a more species-rich vegetation, if further exploited by human use. In addition, the soil is still acidic (low pH). Only a few specialists in flowering plants such as heather can hold their own here, with moss cushions and lichen stands in between, which also manage with very little.

Another type of **marginal landscape** is found in the large parts of Württemberg with calcareous rocks. In the Gaeulandschaften along the middle and upper Neckar and its tributaries these are shell limestones, in the Swabian Alb Jurassic rocks. Where these limestones are covered with clay, e.g. by the flying sediment of loess at lower altitudes or by loam as a solution residue of the limestones, the limestone regions are well suited for arable farming. On slopes, however, or in the immediate vicinity of slope edges along strata and river valleys, the limestone often comes directly to the surface. As the deep soil required for arable farming is lacking here, these areas have often been extensively grazed for centuries, usually as common **land** in common ownership. In contrast to the species-poor Atlantic heath, such **steppe heaths** (according to Gradmann 1898) developed a very large variety of flowering plants in particular. Up to 300 species can be considered characteristic for these rough grasslands and their marginal areas of steppe heath forest in a small area of a few acres.

In purely economic terms, these originally extensive areas can no longer be used today. Natural succession causes scrub to grow up within a few years, primarily with blackthorn (*Prunus spinosa*) and dogwood (*Cornus sanguinea*). Within a few decades, a forest develops which shades the light-demanding species-rich ground vegetation so much that it quickly disappears. It is replaced by a lower species diversity of "common" forest species, i.e. species that are also widespread elsewhere.

As in other protected habitats, it is above all the "beautifully flowering" species that are preferentially protected within the represented flora. In the Württemberg **calcareous grasslands**, these are mainly the **orchid** family (Orchidaceae) with about twenty species and the **gentian** genera with a total of about five species (Fig. 7.2). Although both target groups of nature conservation often occur together, for example in the same slope area, they sometimes have different to opposite **site requirements**. Gentians of the genera *Gentianella* and *Gentianopsis* have coarse leaves and shoots and are thus well protected against the trampling of grazing animals. A powerful ingredient, which in the form of gentian schnapps is quite appreciated by the corresponding clientele, protects the gentians to a large extent from browsing by grazing flocks of sheep. Gentians therefore thrive well wherever grazing is very hard on many other plant species, and thus have advantages within the site competition of species. Orchids, on the other hand, are

Fig. 7.2 Maintenance competitors in rough grasslands: (**a**) German *gentian* (*Gentianella germanica*) and (**b**) the orchid *Ophrys sphegodes*, both attractive target species for nature conservation. (Own recordings)

delicate and resent being stepped on by sheep hooves. They also taste quite good.

In the context of nature conservation, the question of which site requirements which species have is not as academic as it seems. If one actually continues the centuries-old tradition of **sheep grazing** in the calcareous grasslands, one will tend to get a gentian grassland, supplemented by prickly plant species that are also avoided by cattle, such as the silver thistle (*Carlina acaulis*). However, **maintenance** by sheep is usually no longer sufficient and must be supplemented by mowing of the areas (Fig. 7.3). This **mowing** leads to a decrease in gentians and an increase in orchids (Orchidaceae). This is particularly true if – as is often observed – mowing is carried out as early as August, i.e. at a time when the autumn gentians have only just begun to flower. It seems paradoxical when a nature lover is threatened with criminal prosecution for picking a single gentian flower, but a landscape manager mows down hundreds of them in one morning. Nature conservation must therefore decide which species or combination of species it wishes to protect: orchids or gentians. For both together there is no optimal protection on the same area.

Fig. 7.3 On steep slopes, after sheep grazing has been abandoned, the brush cutter is used to cut back the woody plants. The work is lengthy and laborious, thus relatively expensive. It is often carried out on behalf of private landscape managers by the lower nature conservation authority at the district office. (Own recording)

In this context, a (pre-)judgement that is still widespread in nature conservation circles must be cleared up: **Biology**, as a natural science, is not in a position to make a decision in such conflicts of objectives. Behind the goals of nature conservation there are always **value judgements.** As is well known, these cannot be derived from the natural sciences, which always show how something is or functions, but not how something should be. Such value judgements are always *a priori* subordinate to scientific rationality, to take up again the expression of Immanuel Kant already quoted. This means, for example, that biology as a science cannot say whether biodiversity is a desirable goal. If, however, such a value attitude exists a *priori* and remote from the natural sciences, scientific research can certainly suggest criteria and possibilities for implementation.

Conflicting Goals Also Exist in Faunistic Nature Conservation

The protection of large and small owls.

The reintroduction of our largest owl species, the eagle owl (*Bubo bubo*), can be considered a particular success for nature conservation. At least three pairs bred in the immediate vicinity of Heidelberg in spring 2018.

An unforeseen side effect was the decline of the smaller owl species that had previously settled in the area. The previously frequent tawny owls are almost no longer heard calling, long-eared owls at least much rarer. Declines in barn owls are also attributed to the eagle owl. The fact that it has also killed a Little Owl can be considered certain, as an eagle owl pellet was found next to the remains of such a bird.

That the eagle owl also occasionally cranks the peregrine falcon, which is also considered worthy of protection, has also been known for a long time.

However, whether nature conservation under these circumstances can protect at all what it tries to protect is occasionally doubted in the literature:

> Wherever nature conservation now intervenes, whether it 'maintains' the areas or 'leaves them to their own devices': in both cases, the protection alters the vegetation worthy of protection, and, according to all experience, in unforeseen ways. (Hard 1997, p. 568; summarized and also from a more critical perspective: Wilson 2014).

Today, the official nature conservation is recognizably trying to plan for site diversity when redesigning habitats. Thus, packs of quarry stone or old wood are erected, laced together by a kind of cage made of wire mesh. Or timbers with holes of varying thickness are provided as habitat for wild bees. Or nesting aids for songbirds or bats are hung up. All these offers are certainly well-intentioned and may well be important for the survival of individual species.

However, almost all such structural aids to life are subject to a high degree of standardisation, both in their design and in the production process. This only really becomes apparent to the observer when he visits different newly created biotopes one after the other. With this ever-same "individualization" of small structures, however, the traditional diversity of use that is supposed to replace them is precisely not achieved. The basic problem of modern societies is that diversity is always talked about, even in the awareness of a shortcoming, but exactly the opposite takes place. It is not only nature-protected areas that are being standardised, but simply every social subsystem: the production process as such, abstract systems such as administration and the modularised education of our students, right down to concrete projects such as the construction of schoolyards and children's playgrounds. When visiting these, the same boredom spreads as in a large fashion department store when looking at the displays of the different fashion labels: Behind a superficial variety hides the always same or yet similar of a standardized merchandise, which knows the change mainly along the time axis in the succession of different fashion preferences. This enormous tendency towards standardisation also applies to maintenance measures in nature reserves. There, the same maintenance schedule is laid out over the entire protected area with mowing within, say, 2 weeks in August. Originally, this may have been a public open space that was accessible to many users. There, in addition to cattle and sheep, goats and possibly a donkey were used for grazing: animals each with different behaviours in the terrain and with different foraging preferences. Livestock husbandry also had a very different effect along the droving paths and at the stockyards. Perhaps, however, grass and herbs were able to grow higher and denser in early summer in a suitable place despite grazing, and challenged a

Fig. 7.4 Attempt to imitate the traditionally diverse agricultural use by differentiating mowing dates in landscape management (Kissinger Heide near Augsburg). (Own recording)

poor villager to mow with a scythe in order to gain some hay for the winter. In a special situation of scarcity, perhaps a small field was also once created in a deep spot for a summer, which fell fallow again the following year. This traditional diversity of use is the real foundation of the richness of species that has survived. It can hardly be imitated by standardised maintenance (Fig. 7.4).

Yellow Gentian

The following incident actually happened; however, in order not to discredit anyone, no names of places or persons are mentioned.

The **yellow gentian** (*Gentiana lutea*) is generally considered to be a (middle) mountain plant, usually found at altitudes above 500 m NN. It is the largest gentian species with thick fleshy roots, which are also used for the production of gentian schnapps. The plant is not endangered in Baden-Württemberg as a whole. Its main distribution areas are the

Black Forest, the Swabian Alb and the foothills of the Alps. Originally, the species was also widespread in the shell limestone slabs of the Gäulandschaften along the Tauber in the north of the state at altitudes of around 200 m NN, but has disappeared there (Sebald et al. 1996, vol. 5, p. 24).

Now, a few years ago, a single plant was discovered northwest of Stuttgart in a nutrient-poor grassland also little more than 200 m above sea level. The local section of NABU ensured that a rough fence of wooden poles was quickly erected to protect the plant from maintenance operations (mowing down).

On the website of the local NABU the sensational find is presented rather hidden. The side blow against a presumed "nature improver" who might have settled the plant here is recognizably polemic in character. The authors obviously do not like a deliberate change of nature at all.

On the other hand, it could also be that the plant, whose seeds are widely dispersed by the wind, has settled in the new location by itself after all. In this – and only in this – case, conservationists are ready to celebrate this as a great success.

The question of whether the yellow gentian has spread to this location itself or has been settled there by someone else can hardly be answered. The fact that the species existed in larger numbers in the climatically comparable Tauber region until recent times speaks in favour of an autonomous settlement. In favour of an artificial settlement is the distinct uniqueness and isolation of the site compared to those known so far. This plant does not seem to have occurred anywhere in the central Neckar region around Stuttgart in recent times.

Provided that the question of the "artificiality" of the new site could be clarified, the consequences would be clear. In the case of a targeted settlement, the individual of the yellow gentian at the new site would be torn out and this would be considered a further maintenance measure in the nature reserve. In the case of a spontaneous resettlement, on the other hand, NABU would approach a wider public and be duly celebrated for this success: itself and the yellow gentian as a protected species.

The protection by fencing off the Yellow Gentian on the one hand, as well as the polemics against a possible planting on the other hand, betray the aforementioned uncertainty of the organised nature conservationists as to whether the new location is really "natural".

Processes of nature that lead, for example, to the spontaneous settlement of new (albeit native) species are considered worthy of protection in nature conservation. "What nature does", one could say in reference to an old Protestant hymn, "is well done". Nature, regardless of what substantively occurs in it and through it, is considered good per se. It is a quasi-religious superinstance whose properties and behaviour are not to be further evaluated in terms of content.

To the outsider and those unfamiliar with nature conservation, it is not automatically clear why the same circumstance (new location of the yellow gentian) should be good in one case (self-established) and bad in the other (artificially established). After all, we are dealing with the same end state.

The Nature Conservancy is not prepared to evaluate the content of an existing condition (yellow gentian with a new location) on its own merits. Possible settlements here would be:

1. The yellow gentian enriches the species spectrum of the nature reserve XY, increases the local biodiversity.
2. The yellow gentian does not belong in the central Neckar region and must disappear from there again, i.e. be uprooted.
3. Whether there is a yellow gentian at the site is completely irrelevant.

A decision on whether to have the Yellow Gentian at the new location would have to be made purely on the basis of content, without reference to the question of whether someone has located it there (and if so, who).

In philosophy, the moral valuation of a circumstance from the "natural" Thus-Being is seen as a naturalistic fallacy. This is to say that no moral justification can yet be derived from Thus-Being as Thus-Being-Should. The substantive evaluation must therefore still follow (Fig. 7.5).

Moreover, is it really necessary to add that in the cultural landscape all nature is also humanly constituted anyway?

Fig. 7.5 The controversial object: a very vital, almost metre-high yellow gentian surrounded by young growth. The right plant in the wrong location? (Own recording)

8

The Construction of Natural Equilibria: Ideal Starting Point of the Demand for Nature Conservation

Natural equilibrium is a concept in scientific ecology that became firmly anchored in the public consciousness of politicians and the media with the onset of environmental protection around 1970. This remained and still remains the case, even though scientific ecology has already largely abandoned the postulate of equilibrium or at least critically questioned it (for a summary: Jax 1994; Wilson 2014).

Perhaps apart from cyclical fluctuations, which do not, however, call into question the fundamental equilibrium as a whole, **equilibrium concepts** make nature appear as something static, diametrically opposed to the dynamic of human economy and culture. Any change therefore tends to be perceived as negative. The fact that a persistent deterioration of natural conditions perceived in this way appears to correlate with the perceived deterioration of the other living conditions of an aging conservation clientele is what makes it so compelling to most conservationists. However, the "everything was better in the past" also clouds

K.-D. Hupke, *Nature Conservation*,
https://doi.org/10.1007/978-3-662-66159-8_8

the view of the gradual improvements that should also be part of this book: for example, for the increasing immigration of many species into settlement areas and the resulting increase in biodiversity of the urban habitat, or for the successes of large animal conservation in Central Europe (cf. Chap. 19).

In principle, nature is always subject to **change.** The climate, for example, is constantly changing even without any human influence, and all other environmental factors change with it. Neither the distribution areas of plants and animals remain static, nor the number of individuals in populations constant.

Basically, **human interventions** in biotopes and ecosystems merely add to the natural factors that already existed before and continue to exist beyond them. Insofar as human intervention does not lead to total destruction, there is no reason to separate it from natural influences in terms of its intrinsic nature and impact.

This should also be the only reason to speak of a (dis)**disturbance of the natural balance:** If this influence should be so great that in fact nothing comparable will in the foreseeable future take the place of the old and thus destroyed. But such a **total destruction** is also given by volcanic eruptions or by meteorite impacts. The only difference is that under natural conditions such total destructions are rarer by several orders of magnitude, which in turn makes it possible for evolution to react appropriately on a geological time scale.

Figure 8.1 illustrates that there are different ideas about natural equilibria, which are difficult to grasp empirically, and about the resilience attributed to natural systems in the face of external intervention.

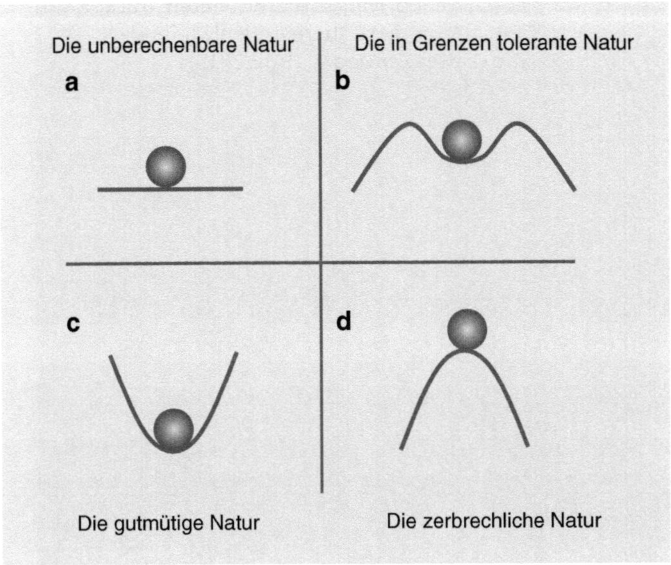

Fig. 8.1 Different ideal-typical conceptions of "balance" in nature. (**a**) the unpredictable nature. (**b**) the nature tolerant within limits. (**c**) the good-natured nature. (**d**) the fragile nature. (Fiene 2013, p. 85, after Kuckartz and Grunenberg 2002)

9

Help for Endangered Species? Red Lists and Endangerment Categories

The blue-winged sand cricket (*Sphingonotus caerulans*) is a field cricket of which there is a local occurrence in the track apron of Stuttgart 21 in the gravel bodies between the tracks. It is specially protected under the Federal Nature Conservation Act and is included on the Red List both for Baden-Württemberg and for Germany as a whole. Its population in Central Europe has been on a downward trend for the past 100 years. So is this alone reason enough to forego the expansion of the former track apron as part of Stuttgart 21 from a nature conservation point of view?

There may be a total of 50,000 animal species in Germany. Insects are particularly **rich in** species, especially Hymenoptera (wasps, bees, ants, etc.) as well as Diptera (mosquitoes and flies), beetles and butterflies. Among the vertebrates, birds with about 300 species can be considered particularly diverse, in addition to fish. More than 3000 species of flowering plants are represented.

K.-D. Hupke, *Nature Conservation*,
https://doi.org/10.1007/978-3-662-66159-8_9

Among these thousands of species, most are rare and always threatened by this fact alone. The individual species may be rare in itself (this is also part of our Red List definition). But since there are thousands of species in total, they are again numerous in their sum total. You only have to study a habitat long enough to find some of the **endangered** species, no matter how remote and **species-poor** it may be. This is usually done with the help of Red Lists, which can be used to objectify the **worthiness of** a species to be **protected.** At the very least, the Red List is considered to be expertly identified and was developed for the **assessment of habitats** (nature conservation assessment).

It is therefore self-evident that proving the worthiness of protection on the basis of certain individual species cannot be an appropriate means of highlighting the importance of a biotope for nature conservation. Rather, one would have to look for **indicator species** whose existence would speak for a biotope worthy of protection as a whole. But even then, one would have to agree on further criteria of worthiness of protection, such as pristine nature, species diversity, usability or scenic beauty.

But what do **Red Lists** for the detection of rare or endangered animal and plant species actually consist of and, above all, how do they come about? Red Lists exist at different spatial levels. On the one hand, there are international lists compiled by the International Union for Conservation of Nature (IUCN). These can best be regarded as absolute benchmarks for the endangerment of living beings, as they assess the **global threat.** Of regional importance are the national Red Lists, which are supplemented at a further sublevel by the Red Lists of the German Länder, for example. These attempt to reflect the threat at the respective regional level.

According to the **degree of endangerment,** the Red Lists are also numerically graded, in Germany there are the categories:

- **0** Extinct or lost.
- **1** Threatened with extinction.
- **2** Critically endangered.
- **3** Endangered.
- **4** Potentially endangered (state level only).

It goes without saying that the regional lists and the global list of the IUCN must inevitably differ from each other. For example, many species are rare in Germany only because they reach the **limit of their natural range** here. This applies, for example, to steppe species such as feather grasses (*Stipa* spec.), whose isolated occurrences are strictly protected here, but which are much more common in the natural steppe areas of, for example, Ukraine and, above all, Mongolia, which is used more extensively for agricultural purposes. The same applies to Mediterranean species such as many meadow orchids, whose distribution to the north often extends only as far as south-west Germany or Thuringia. And this applies to northern European species such as the marsh spurge, a species of the genus *Rhododendron* with a distribution focus in Scandinavia, which in Germany is restricted to a few moorland sites, especially in Mecklenburg-Western Pomerania. Climate change brings additional dynamics, especially at the range boundaries, with the result that many orchids are spreading in Germany, while the marsh spurge has almost completely disappeared from southern Germany (except for a few sites in Bavaria) over the last few decades.

Many species also have a huge range, for example many "birds of prey" such as the *peregrine* falcon (*Falco peregrinus*) and the osprey (*Pandion haliaetus*), which are virtually **cosmopolitan.** Such species are regularly rare in their ranges and therefore play a major role on regional Red Lists, but are less threatened globally. In the case of other species, their distribution is restricted to a rather small area, within which, however, they are relatively common. For example, much of the Red Kite's population is concentrated in Germany, where, however, the species is the most widespread and common hawk (Accipitridae; birds of prey, minus falcons) after the Common Buzzard (*Buteo buteo*). The copper beech is the most widespread forest tree in Germany by nature and also to a large extent in reality. Outside Central Europe, however, the species is restricted to a few neighbouring mountain areas and may be considered threatened in large parts of Europe.

Species with a very small distribution, so-called **local endemics,** are particularly threatened. In particular, old and spatially isolated habitats have many such local endemics, for example on islands and in mountains. Germany has only very few endemic species. The best known of these is the Lake Constance forget-me-not (*Myosotis rehsteineri*, Fig. 9.1), which, however, is not strictly speaking endemic to Germany, but to the gravel banks of a few Alpine waters, especially Lake Constance.

However, species that are quite widespread as cultivated plants are also considered threatened in their endangered wild occurrences. One example is the wild grapevine (Fig. 9.2). Another criticism of the Red List concept is that it is based on distinct, strictly distinguishable species. However, **the concept of species** originates from the idea of creation and has been scientifically outdated at the latest since Charles Darwin's theory of evolution. There are no species separated from each other by genetic leaps, but only populations with a given gene pool, which are connected to

Fig. 9.1 The Lake Constance forget-me-not (*Myosotis rehsteineri*) is found only on sparsely vegetated gravel beaches of some pre-alpine lakes. (© Tragopogon; CC-by-SA-3.0)

other populations both spatially and developmentally by fluid transitions. The pure species concept on which Red Lists are based runs the risk of protecting only these species or, where appropriate, clearly distinguishable subspecies, but of underestimating the genetic and conservation value of local populations. Nevertheless, in the absence of alternatives, Red Lists remain a necessary concept for conservation practice icic diversity.

Fig. 9.2 The wild grapevine (*Vitis vinifera sylvestris*) is one of the rarest plant species in Central Europe. Most of the remaining wild stocks (a few dozen specimens) are found on the Ketsch Rhine Island (south of Mannheim), which was created by the straightening of the Rhine and covers only about 500 ha. It is presumably a cultural relic of Roman times, when viticulture was established north of the Alps. Through the excretion of grape seeds by birds, the plant also came to the alluvial forests, where it found such favourable conditions that it was still extremely common along the Upper Rhine at the beginning of the nineteenth century (Sebald et al. 1992, vol. 4, pp. 127–130). The decline has not yet been completely clarified, but is probably related to the changes in water regime, flow dynamics and vegetation structure associated with the straightening and navigability of the river. In the picture, the leaves (about in the middle of the picture) as well as the ripe grapes (further to the left) are clearly visible at about 6 m tree height

Securing Bavarian Spoonwort and Other Endangered Species

In future, the Bavarian Nature Conservation Association (BN) will devote special attention to one plant species: the "Bavarian spoonwort" (*Cochlearia bavarica*). It only occurs in two areas in Bavaria and is a so-called "endemic", for whose conservation Bavaria bears global responsibility. As part of the Federal Programme for Biological Diversity, the occurrences of this species are to be recorded in the biodiversity project "Löffelkraut & Co." until 2016 and secured through various measures. The focus is on plant habitats such as near-natural lime springs, little-impacted spring brooks and stream headwaters. The BN receives about €450,000 from the Federal Programme on Biological Diversity [...] for securing the spring habitats and protecting the endangered Bavarian Spoonwort and other endemics.

(Bundesamt für Naturschutz 2013a, p. 27).

Petite Plate Snail Stops Major Project

The slug (*Anisus vorticulus*) is only five millimetres in size. Nevertheless, it is causing a project worth millions to totter. The construction of a logistics park in Bergedorf will be delayed for years because of the tiny snail. This was confirmed by the responsible district office in an interview with NDR 90.3. Actually, the logistics park between the A 25 and the street Curslacker Neuer Deich should be ready for the International Building Exhibition in 2013. However, after the discovery of the snail threatened with extinction, there is now talk of 2016–3 years later – at the earliest. According to the district office, an attempt will be made in September of this year to relocate some of the snails to areas in the Vier- und Marschlanden.

Tellerschnecken are to be Relocated

Whether the relocation is successful will be shown according to the information in 2 years. Only then can the remaining animals be relocated. And the development of the site will have to wait until then. Should the relocation fail, further delays are imminent. So far, the district office has not given any information about the costs. At the moment, preliminary investigations are still underway, it said. In addition, the district office has commissioned several surveys, also to examine alternative access routes to the area. This is because the ornamental snail (*Anisus vorticulus*) has settled in the very spot where the driveway is planned.

(https://www.ndr.de/regional/hamburg/bergedorf157.html).

Author's question: How useful is relocation if the habitat in question is destroyed?

10

Of Birds and of Butterflies: How Nature Conservation Distributes Its Sympathies

Generally speaking, birds are probably among the most **likeable creatures** in the animal kingdom (cf. Fig. 10.1). We enjoy their often bright colours. We share in their departures in autumn and their return in spring, in the building of their nests and the rearing of their young, we scatter food for them in winter or even build our own bird feeders. Song is considered beautiful in most species, although its sole purpose in evolutionary history was probably to delineate territories or attract a sexual partner. Most conservation organizations dedicated to a specific group of animals are for bird conservation. *Bird watching* is one of the most common outdoor activities and special travel motifs, especially in the Anglo-Saxon world.

Among the arthropods, which are not closely related to birds, **butterflies** can hope for similar sympathies. There is a superficial similarity to birds in that **butterflies** can also fly and are often brightly coloured. Possibly, these two characteristics are also the basis for the respective sympathy. Many people may be disgusted by caterpillars, but no one is

K.-D. Hupke, *Nature Conservation*,
https://doi.org/10.1007/978-3-662-66159-8_10

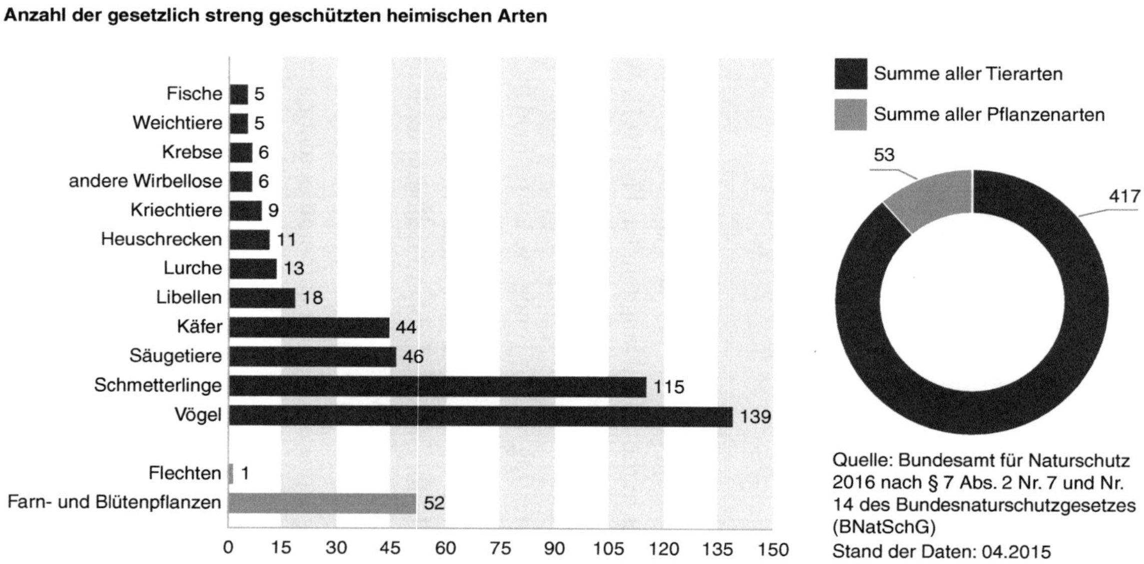

Fig. 10.1 Birds and butterflies: preferred by official nature conservation. (Source: BfN, Homepage: www.bfn.de/23521.html, ges. 5.1.2017)

disgusted by a butterfly. Butterflies remind us of sunny summer days and we often associate them with the equally sympathetic flowers on whose blossoms they suck nectar. Their darting flight does not seem as purposeful as the flight of flies, dragonflies, bees or wasps, and is therefore hardly ever perceived as aggression. Moreover, every child knows that butterflies cannot sting. This is also true for dragonflies, but in this case it is not generally known.

Other arthropods are not met with comparable sympathy. This is especially true for **spiders** (Araneae). People know that they eat other animals, especially insects, and have respect for them. In addition, they often do not move for a long time, and if they do, then hastily. This again makes sense in the context of evolutionary history: Either you try to remain motionless in the face of a larger animal, as humans probably are, and rely on your camouflage. Or you run away, but then as fast as possible. The fact that spiders also produce very beautiful things, namely the spider webs, unsurpassed in their geometric pattern, is not perceived as intensively as it actually deserves. I have met female students—and not only in individual cases – who in all seriousness refused a study excursion to a tropical country on the grounds that there would "certainly be large spiders" there. In order to justify the danger, reference is regularly made to the venomousness of some spider species. But hardly anyone knows of a case where anyone has been seriously harmed by a spider bite, except through their own disgust. Even the bite of the legendary black widow (esp.: *Latrodectus mactans*) is only fatal in exceptional cases, for example if the person bitten has a special allergy.

Of course, one could now cite the widespread presence of large hairy **tarantulas** (Theraphosidae) in the rooms of young people and adults as counter-evidence to the lack of human sympathy towards spiders. But, paradoxical as it may sound, the popularity of tarantulas feeds primarily on the disgust of each other and can also be seen as a means of self-dramatization. It is no coincidence that tarantulas were initially particularly widespread among punks: as anti-sympathy carriers, like rats.

One could easily dismiss these sympathies or antipathies as irrelevant. But on the one hand, the "nice" species determine both the motivation and the substantive focus of nature conservation. On the other hand, broad and media attention can only be achieved with sympathetic species. The **WWF**, for example, as the largest international nature conservation organisation, is currently working very hard with the polar bear as a popular figure, which is also a good way of demonstrating the expected consequences of climate change. Parallel to this, and historically preceding it, the fascinating big cats have been a particular focus for the WWF (Chap. 3). The hype surrounding the polar bear cub "Knut" alone, as well as the decades-long advertising mascot of a mineral oil company, show that such sympathy values are not limited to conservationists, but aim for the middle of society (where nature conservation was actually never really located and often felt misunderstood). While in the case of the bear it is the cuddly-tapery that arouses sympathy as well as the large head that activates the childlike schema, in the case of the cat it is the beautiful fur markings (which are inconspicuous in the natural vegetation and rather serve to camouflage the predator).

Even when designating nature reserves, orchids (Orchidaceae), gentians or beautifully flowering and conspicuous individual species such as the marsh hellebore (*Gladiolus palustris*) or daffodils are regularly mentioned among plants (Figs. 10.2 and 20.9). The mostly inconspicuous and not very colourful crucifers (Brassicaceae) and umbellifers (Apiaceae), on the other hand, almost do not appear, although both are very species-rich in the Central European flora and contain many threatened species.

Fig. 10.2 "Beautiful" target species of nature conservation with striking colours or strong colour contrasts. (**a**) Lady's slipper orchid (*Cypripedium calceolus*). (**b**) Stemless gentian (*Gentiana kochiana*). (**c**) Fire salamander (*Salamandra salamandra*) (the specimen shown is, incidentally, a deceptively good imitation as a child's toy). (**d**) Admiral butterfly (*Vanessa atalanta*) on a butterfly bush. (Own recordings)

11

What Endangers Nature?

An at least potential **endangerment of nature by humans** can be imagined on different levels.

For as long as humans have existed, i.e. for at least a few 100,000 years, humans have been using nature and thereby changing it. At first, like other animal and plant species, he was more a part of nature and did not yet change it so spectacularly that one would have to speak of a threat. However, this already changed with the introduction of **fire**, which possibly dates back to the middle Pleistocene, i.e. to about 500,000 or even 700,000 years ago (Goren-Inbar and Speth 2004). At that time, a landscape change probably occurred first in Africa, leading to a transformation of mainly tropical deciduous forests into **savannas** and grasslands. To a contemporary critical observer, this process, which presumably took place slowly in historical time scales, could well have appeared as the destruction of nature (whereas the results of this landscape change that are still visible today are received on safaris as nature experiences). After all, this

K.-D. Hupke, *Nature Conservation*,
https://doi.org/10.1007/978-3-662-66159-8_11

process probably took place slowly enough and was long enough ago for the animal world to be able to adapt evolutionarily and for us to encounter a large animal fauna in Africa today that, although causally (at least partially) anthropogenic, appears as untouched nature.

The effect of fire was supplemented by **hunting.** This could and can lead to a decimation of certain animal species, even to extinction, in a large animal world that is not adapted to this hunting situation through increased reproductivity. These early influences also lie so far back in Africa that they no longer appear to us today, if they can be reconstructed at all, as an impairment of nature.

In all other continents, the immigration and proliferation of modern humans at different times probably brought about **Holocene overkill** (Martin and Wright 1967), the extinction of most large mammal species. There is no question that, as far as large mammals and flightless large birds **are concerned**, the massive **extinction of species** has already taken place. To put it bluntly, 5000–8000 years ago, there were so few species left on almost all continents that the extinction of the remaining populations would be almost irrelevant.

The spread of **agriculture** over the past 10,000 years, initially in the river lowlands and steppe areas of what is now the Middle East, then in South and East Asia, in Africa and Europe, and in historical times in North America and Australia, set further accents in the area by reducing near-natural habitats to the remaining regions that were often unsuitable for agriculture. In recent decades, due to the increasing global demand for food, industrial raw materials and, above all, energy substances, many of these unsuitable zones have been further developed through irrigation or improved agricultural techniques, such as the steppe and desert lowlands of the Old World and the wet savannahs and semi-evergreen forests of South America.

In addition to the worldwide expansion of agricultural areas at the expense of natural forests and grasslands, the **intensification of agricultural use has** recently played a major role in the decline of animal and plant species. The increased use of fertilisers (including natural fertilisers) and pesticides often jeopardises the balance that has evolved over thousands of years between farmed species and wild species that have migrated into the farmland. In the old-established areas of Europe and large parts of Asia, this factor is currently the main threat to biodiversity.

Modern **industrial societies are** nevertheless taking up more and more space for housing, leisure activities, traffic and transport, while the population is hardly increasing any more. It is true that urban nature has been and continues to be enormously enriched by an increasing number of immigrating animal and plant species (Chap. 15). However, this only applies to open spaces and open areas within urban areas. Housing densification or more efficient land use also threaten nature in the city.

The achievements of a modern **infrastructure** such as wind turbines, reservoirs, road constructions and airports lead to a further loss of land, especially in rural areas far away from settlements. On the other hand, these facilities also directly threaten wildlife, for example when birds of prey are killed by the rotors of wind turbines or migratory animals are run over on the roads. Linear infrastructures such as roads in particular have the additional effect of preventing or at least restricting migrations of flightless animal species. The increasing **densification of traffic and settlement areas** has been and continues to be recognised as one of **the** main obstacles to nature conservation. While in the 1970s the discussion about the increase of built-up areas determined the situation (Tesdorpf 1984), this problem has eased somewhat. On the one hand, this is due to the realisation that, in view of the increasing remoteness of many agricultural areas from nature and an increased importance of built-up (but not entirely "sealed") settlement areas such as parks and gardens for nature conservation, the discussion in the past was too sweeping. On the other hand, landscape consumption, which is mainly due to the growth of settlements, has been declining somewhat in the industrialised countries in recent decades as a result of lower economic dynamism and demographic stagnation.

All the more attention is being paid to the **islandization of habitats** by major roads (Fig. 11.1). For many animal species, this results in small and very small populations between which there is no longer any exchange. This poses two dangers.

On the one hand, the probability increases that subpopulations with very few individuals gradually die out due to chance. This can be caused by predators such as birds of prey, foxes or martens, but also by extreme weather conditions or an imbalance in the sex ratio. In contrast to a large and

Fig. 11.1 The satellite image from the commercially available google maps shows the immediate vicinity of an intensively used agricultural landscape (right) and a landscape of forest and lawns in the area of the garland-like curved motorway feeder roads, which appears close to nature in detail. The problem, however, is the extreme fragmentation and isolation, which excludes all non-air-mobile species

spatially coherent population, such a **local extinction** can no longer be compensated by immigration. As a result, there is a risk that such an isolated population will gradually disappear altogether.

On the other hand, in a very small population there is also a **reduction in genetic diversity over** time due to the accidental loss of individual genes. Geneticists generally assume that at least 50 individuals are required in the medium term, and at least 500 individuals in the long term, to ensure sufficient genetic diversity within a population.

In the Mannheim area, the local populations of the field hamster (*Cricetus cricetus*) are already characterized by a strong islandization by motorways and federal roads, but also commercial and residential settlements. Several subpopulations have already become extinct as a result. With the participation of the Heidelberg Zoo, a genetic exchange between the remaining isolated occurrences has been carried out for some time, with significant use of released captive-bred animals.

Plants are also subject to this insect effect. However, since many species have seeds that can fly or that are so small that they can be dispersed by the wind (as in the case of many orchids), the flora as a whole is often less affected. But here, too, larger, more or less homogeneous areas per unit area are usually richer in species than smaller ones, as can be seen, for example, in the case of rough grasslands in the central Neckar region.

However, there are also **more indirect threats** that can have a negative impact on nature conservation. An illustrative example of this is the disappearance of the lesser wintergreen (*Pyrola minor*), a – as the name suggests – rather inconspicuous little plant that was still more common decades ago in the lean (= nutrient-poor) pine forests on shifting sands south of Heidelberg. The reason for the decline of this species is seen in the nitrogen input as a result of increasing road traffic (cf. also Fig. 11.2). It is assumed that

Fig. 11.2 The bug orchid *(Anacamptis coriophora)*, a rather inconspicuous native orchid, was scattered over almost the whole of southern and central Germany on rough grasslands about 100 years ago. Today it is restricted to a few sites, all of which are far from large cities or are located west of densely populated areas, such as the Palatinate (west wind zone!). The disappearance is, if one takes these location factors into account, probably due to nitrogen inputs from road traffic via air currents. (Own recording)

the increased nitrogen content led to the wintergreen being displaced by more vigorous plant species of the forest floor. In addition, a greater supply of nutrient salts enables a denser formation of the tree crowns, so that the summer-flowering plants, which unlike the early-flowering plants (Section "Forest Ecosystems") cannot hope for abundant sunlight, are too heavily shaded. We do not know for sure why wintergreen has really disappeared from Heidelberg's forests. We only know that it became increasingly rare and finally disappeared altogether, and it seems reasonable to suspect a direct human environmental influence.

Too Much Nitrogen: EU Admonishes German Government

For years, Germany has been exceeding the limits for the air pollutant nitrogen dioxide (NO_2). It is mainly caused by diesel vehicles. But the federal government refuses to take responsibility.

Not just since today, but since 14 January of this year [i.e.: 2015; d.Vf.], the Federal Government has had it in black and white. On that day, the German Advisory Council on the Environment (SRU) presented its report entitled: "Nitrogen – solution strategies for a pressing environmental problem."

(Source: https://www.dw.com/de/zu-viel-stickstoff-eu-ermahnt-bundesregierung/a-18570067, ges.: 13/1/2017).

Note from the author:

The compulsory introduction of catalytic converters for new cars in 1989 was more than compensated for by the simultaneous shift in passenger car sales towards diesel vehicles with mineral oil (= energy) tax advantages.

Apart from this, however, it must be warned against considering natural areas (without human intervention) to be static. For natural reasons such as succession, climate change and competition between species, there have always been changes in the distribution of probably all plant and animal species. To regard humans alone as dynamic, while nature is always static, is a widespread anthropocentric error (cf. Chap. 8).

The effects of the individual hazard factors are therefore manifold. The **growth of the human population** is accompanied by a technically induced and thus possible increased consumption of food, raw materials, energy sources and land—in other words, an increased consumption of basically almost everything and everyone. And this happens in conjunction with an economic and social system that not only does not set any limits to this increase in consumption as "economic growth", but downright idealizes it and explicitly demands it, even requires it for survival.

In contrast, the targeted killing of animals, for example through hunting, the targeted capture of particularly attractive species for keeping in captivity, or the targeted picking or digging up of beautifully flowering and rare plants account for a smaller proportion, which in individual cases can nevertheless lead to a targeted species becoming rare or completely extinct. The same applies to the direct effects of road traffic (Figs. 11.3 and 11.4).

Fig. 11.3 For slow animal species, roads form barriers to dispersal or can only be crossed with high losses, which can also endanger a population in individual cases. The picture shows a fire salamander that has been run over on a dirt road. (Own recording)

Fig. 11.4 Significantly damaged carapace (upper shell) of a Greek tortoise near the zenith in Montenegro. The accident is likely to have been caused by a traffic accident or farming equipment. The damage is halfway healed, the animal survived the accident. (© Wenzel Halla and Stefan Hupke)

Road Traffic and Turtles

Intuitively, busy roads are generally thought to have a negative impact on reptile populations in their vicinity, but specific studies in this regard are few and far between. US researchers have now documented the influence of road traffic using the California gopher tortoise (*Gopherus agassizii*) as an example.

The researchers found that both encounters with and evidence of the presence of turtles (Testudines) correlated significantly with the amount of traffic in their vicinity. Not only were fewer turtles encountered the busier the roads in the area, but the turtles near busy roads were also significantly smaller and younger on average. This effect was clearly visible up to a distance of 200 m from the traffic arteries.

The researchers conclude from their observations that road traffic affects gopher tortoise populations in several ways. On the one hand, animals are run over, and on the other hand, the reproduction rate of the population decreases, since traffic deaths particularly often affect the large adults and slow specimens that contribute most to reproduction. The impact of the traffic network on gopher tortoises is thus significant and, according to the scientists, is representative of the dangers that road traffic generally poses to low-mobility, slow-reproducing species.

(Minor. Informationsbrief der „Arbeitsgemeinschaft Schildkröten "2013, p. 15).

Roads Make Life Difficult for Wild Animals

For humans, roads are connections—but for nature, they are cuts. According to an extrapolation by researchers, areas without roads still account for 80% of the earth's land area, but they are highly fragmented. Areas were defined as "free of roads" only if they were at least one kilometre away from roads, but also from smaller axes such as forest tracks. A two-kilometre-wide forest between two traffic axes, for example, was therefore not included.

In total, the team led by Pierre Ibisch from the University for Sustainable Development in Eberswalde came up with 600,000 pieces of land lying off the beaten track worldwide – in Germany, at most, there are only a few.

(n. dpa, Stuttgarter Zeitung v. 19.12.2016).

12

"Outlaws" and "Helpers": The Players in Nature Conservation

The following scene is by no means fictional:

Two children of primary school age are romping along a path at the edge of the forest. A younger couple joins them, the man asks the children, definitely with an annoyed undertone: "Say, don't you know that this is a biotope?"

The edge of the forest in question was in no way part of a nature reserve or subject to any other more specific protection. However, some questions inevitably arise in connection with this little incident. First of all, it would have to be asked whether the man would have seen a similar problem if, for example, a larger dog had been frolicking at the edge of the forest. Or whether he would have been annoyed if tracks indicated that a troop of wild boars had been ploughing through the edge of the forest at night. At the very least, the latter would cause far greater damage to vegetation and wildlife than rampaging children.

K.-D. Hupke, *Nature Conservation*,
https://doi.org/10.1007/978-3-662-66159-8_12

Presumably, however, both dog and wild boar would not have disturbed the walker at all. The reason lies solely in the fact that both the dog and the wild boar belong to the area of nature. And the man probably wanted to stay in this area and not be disturbed by playing children (as in Figs. 12.1 and 12.2).

A similar "problem" arises when children catch a frog or a tadpole, which they like to do as "hunters and gatherers". However, this is prohibited because all frogs native to Central Europe are protected. In case of doubt, parents are liable because they have violated their duty of supervision.

Fig. 12.1 Like hands reaching into the void, these tree roots appear to be searching for a foothold in the soil. Probably thousands of playing children have trampled the soil in the surrounding area over the years, thus increasing the surface runoff of rainwater in the pathway slope and thus the soil erosion. The surface in the basal zone of the copper beech has thus been lowered by about a quarter of a metre. As the tree is increasingly losing its stability, it will probably have to be felled in the next few years for safety reasons. This is a nuisance for the largest local nature conservation association. But how should one react (outside a nature reserve) to (justifiably) playing children: with fencing, with bans? (Own recording)

Fig. 12.2 Children playing by the stream: Playing in nature should be regarded as a meaningful experience rather than as destruction of nature. (Own recording)

However, if you ask convinced conservationists about their first **encounters with nature**, memories of captured beetles, frogs, newts or lizards almost always come to light, not infrequently transfigured in retrospect.

I can well remember a larger pond on the edge of the then new housing estate where I spent my childhood. It was a gypsum keuper doline, over the bottom of which loess clay had laid an impermeable cover. On fine spring and summer days almost all the male youth of the settlement waded through the pool, which was only one acre in size and largely shallow, with mother's canning jar in hand, which gradually filled with prey. Most of the captured animals were eventually returned to their habitat by their parents. Many, however, did not survive captivity. Still, the four species of amphibians found there never went extinct. They only disappeared when an avenue of old pear trees nearby was cleared, which had served as a residence for the tree frogs outside their spawning season, but especially when, at the suggestion of conservationists, willows were planted in the pond and at its edge. The trees, which soon became quite large, removed almost all the water from the pond, so that no amphibian brood can grow in it today. The mass occurrence of at least three amphibian species (tree frog (*Hyla arborea),* yellow-bellied toad and pond newt) resulted from a special constellation of tree-free areas in the immediate vicinity of the pond, which protected the water level, and an avenue not too far away, which provided living opportunities for the tree frogs and probably also the pond newts. The fact that the children caught the animals in not inconsiderable numbers was of no consequence. Especially reproductively strong prey animals can be enormously productive if the structures of the habitat are right. But it was not the children who destroyed this constellation, perhaps not even the felling of the old cider pear trees in favour of a settlement extension. The main problem in this case was actually official nature conservation, which could well have imagined a few willows at the otherwise "ugly-bare" waterhole to "liven up" the landscape and failed to recognize that it was precisely the unimpeded incidence of sunlight that offered a special opportunity for the pond newt and toad, both of which avoid shaded waters.

More generally, for children, the motivation of catching animals is based on the particular incentive to engage with nature at an early age. If, in comparison, one observes a grey heron for only half an hour, it will be easy to see that it is by far the better small animal catcher and that catching by children (or also by adult hobby aquarists and terrarists) is not a problem in relation. This only becomes a problem if one is caught up in the **nature-culture dichotomy** (Chap. 35) and sees something fundamentally negative in human intervention. On the contrary, the capture of animals by children shows how much humans are part of nature, just as much as white storks and grey herons, simply because of their natural endowment.

In rough grasslands, annual **mowing** is a necessary maintenance measure to preserve the target biotope. However, as already explained in Chap. 7, this measure comes at an inopportune time, at least for the **late bloomers** involved, if it is carried out as early as August, as is frequently observed. In the dry grasslands of the central Neckar region north of Stuttgart, hundreds of flowers of the fringed gentian *(Gentianopsis ciliata), the German wreath gentian (Gentianella germanica)* or even some of the much rarer cross gentian *(Gentiana cruciata)* are regularly found dried up on the ground. Not to mention the other late-flowering species that are also worthy of protection, such as the Carthusian carnation *(Dianthus carthusianorum)* and the Lime aster (*Aster amellus).* In this context, it seems strange and downright paradoxical that a walker would have committed an offence by picking a single flower the day before. This would still be the case even if he had subsequently taken a flower that had been mown by the official nature conservation service.

In view of the "destruction potential" of otherwise quite appropriate **maintenance measures,** the legal regulations within a nature reserve seem excessive in many respects anyway. This applies in particular to the **prohibition of access** outside the paths. Such a ban may seem sensible in an area as heavily frequented as the Lüneburg Heath, where crash barriers along the paths direct the flow of visitors. But here in particular, the heavy footfall should rather be an incentive to create such areas on a larger scale and thus to spread the footfall to a less harmful level.

Particularly in connection with the **impact on** vegetation, it must be pointed out that access to nature tends to be made more difficult for people who are particularly interested in nature. An important motivational and multiplier function of nature reserves for nature conservation is thus also made more difficult or even lost completely. On the other hand, one has to observe a herd of sheep once, how they deal with the existing vegetation with their sharp-edged hooves – in the context of a "maintenance measure". Of course, this is not to claim that human tread would be comparable "care". But the comparison clearly puts the possible damage into perspective.

Committed nature conservationists will object that the prohibition of walking on the land serves not only to avoid foot traffic but also to prevent other forms of damage: it prevents animals from being disturbed, for example, and in particular protects ground-nesting birds. However, the historical-spatial comparison teaches that there is hardly any animal species that avoids humans "by nature", unless they pursue them. The **cultural fugitive concept of** classical conservation biology has finally lost its justification. And even ground-nesting birds are much more likely to be threatened by dogs than by isolated hikers or walkers per se.

People who visit nature for sporting or other **leisure interests** are also a thorn in the side of many conservationists. These include climbers, who seek out the relatively isolated natural rocks outside the high mountains or quarry walls for their exercises. However, these are also the preferred nesting sites for preferred species of nature conservation such as the eagle owl and the *peregrine* falcon *(Falco peregrinus)*. In many cases, a seasonal **ban on climbing** during the animals' breeding season could provide a remedy. However, the characteristic rock vegetation with a large number of species that are particularly adapted to the site conditions such as lack of soil crumb, strong sunlight and dryness, e.g. houseleek species (*Sempervivum* spec.), continues to be impaired. In this case, climbing represents almost the only conceivable direct encroachment on an extreme site that is otherwise difficult to reach by humans and hardly usable economically.

In a similar way, smaller to medium-sized rivers are increasingly used for canoeing. Commercial operators, in south-western Germany for example on the upper Danube,

Enz, Kocher and Jagst, rent canoes, take them back a few kilometres downstream and arrange for the return transport of the canoeists to their parked cars. In this case, too, the concerns of nature conservationists were and are great, and here, too, seasonal **canoe bans** during the breeding period of waterfowl as well as a ban on navigation below a certain water level have provided some relief. However, such restrictions (in winter and during high water the rivers cannot be navigated anyway) result in a shortening of the season and thus in an economically often only unprofitable short period of use. Thus, overly strict conditions often result in restrictions on use.

The main potential threat, on the other hand, is not seen so much in individuals, but in structural threats such as the standardisation of the field (Chap. 2) and the increasing use of agricultural chemicals (Roundup/glyphosate against "weeds", neonicotinoids against insect infestations). The "insect mortality" that has been increasingly discussed in recent years, even in the mass media, is attributed to the sum of the above-mentioned influences. And there are certainly warning signs in this direction, which, however, require further confirmation of scientific research that is also critical of its own methods. The malaise traps, which are mostly used, are particularly effective against hymenoptera, i.e. bees, wasps and (fusiform) wasps: Bees, wasps and (flying) ants). The "dying of the bees" thus finds its background. To be more precise, it is particularly a decline in the very species-rich wild bees (in Germany more than 500 species in total). However, the increasing number of hobby beekeepers with their many unregistered colonies could also contribute to this. The honey bee *(Apis mellifera), which is* very competitive because it is very rich in individuals and bred for pure honey yield, poses a potential threat to wild bees (Steffan-Dewenter and Tscharntke 2000; Geldmann and González-Varo 2018). Beekeepers also like to set up bee boxes, especially in the vicinity of nature reserves, as the flowering period there, which is often almost year-round, promises an optimal yield over the entire year. As the reaction to the formally successful Bavarian "Volksbegehren Artenvielfalt – Rettet die Bienen" (petition for a referendum on biodiversity – save the bees) of spring 2019 has also shown, in the media context no distinction is usually made between wild and honey bees, and beekeeping receives support in a context in which it could also be seen as (co-)causing damage. Apart from wild bees, honey bees also compete for food with other nectar- and pollen-collecting insects (e.g. butterflies).

The conservationist sees himself above all as the counterfigure to these at least potential endangerers and harmers of nature. Here, however, it is first necessary to define who is to be considered a **nature conservationist in** the first place. First of all, those who are officially entrusted with this task: within the district offices, the regional councils and ministries, the relevant state institutions, the Federal Agency for Nature Conservation. However, nature conservation thrives not only on offices and official functions, but above all on the **voluntary commitment of** individuals (Fig. 12.3) and associations, among which the **NABU** (Nature and **Biodiversity Conservation Union)** and the **BUND** (**Friends** of the Earth Germany) are particularly prominent in Germany. As a rule, volunteers also hold the office of district nature conservation officer within the framework of the lower nature conservation authority.

It is precisely this voluntary work that is in no way to be disparaged in its idealistic approach. Nevertheless, one cannot avoid noting that those entrusted with this task are sometimes granted exclusive rights in return. These include, for example, largely unrestricted **access to nature**

Fig. 12.3 Knowledge as a prerequisite for nature conservation: A volunteer conservationist has created boards about the endangered animal and plant species found in the nature reserve. (Own recording)

conservation areas, even off the beaten track, for reasons of control. Where a nature reserve is provided with a fence and an entrance door (Figs. 12.4 and 12.5), which may well be the case especially at the sites of beautifully flowering plants (e.g. Lady's Slipper Orchid, *(Cypripedium calceolus),* the nature conservation officer is literally given "key rights". Such a key may not be passed on under any circumstances; it is tied to the individual. It is precisely the comparison of these privileged accesses with the barriers to access of the ordinary citizen that indicates a power imbalance that conservationists regularly emphasise by virtue of office and in principle in particular. Such an office creates quasi-exclusive access to nature. Since nature conservationists are often (but by no means always) joined by a rather introverted nature that interprets nature as a counter-world to crowds of people and human societies (cf. Chap. 1), this exclusivity directly results in personal development opportunities that

Fig. 12.4 Recreation-seekers are often denied access to protected areas. (Own recording)

Fig. 12.5 In this example, even the explanatory board, which is actually intended for the general public, is "secured" behind the wire mesh fence, which does not necessarily serve its legibility. If it does, the nature-interested person who remains outside is at least given an idea of what he might see inside the fence by the picture board. (Own recording)

otherwise only a large landowner, a nobleman with a large private park or a multi-millionaire who owns his own island possess. However, since the holders of such (honorary) offices are usually "only" small- to middle-class senior councillors, university lecturers or freelancers, such an increase in power and competence is already enormous.

Knowledge of the occurrences, especially of the beautifully flowering and rare preferred species of nature conservation such as orchids, is passed on as a "secret" and only revealed to a select few. The argument is usually that one does not want to threaten the occurrence. But who benefits from such an occurrence that (almost) nobody knows about? Again, exclusive access to nature is created and a secret society of conspirators is constituted. The ethos behind this is elitist and neither compatible with Kant's categorical imperative nor with the approach to consciousness of modern constitutions and concepts of man.

The question arises again here:

For whom do we want to protect nature? Who should it serve, who should have direct access, who should be allowed to look at it?

A study conducted a few years ago (Kuckartz and Rädiker 2012) confirms both the external impression and previous research that **the acceptance of nature conservation** (in this case its main goal: biodiversity conservation) increases with increasing social position, with increasing income and with increasing formal education. Of course, knowledge is required to recognise the value of diversity in nature. But this is not automatically taught in school, even less so today than a few decades ago. Maslow's **pyramid of needs** (reprinted in 2013) might be more helpful in clarifying the connection here. In the sense of a tiered pyramid of needs, the desire for an intact nature and environment ranks quite high and thus appears to be of lower priority, at least for lower income groups. Basic needs of people with low disposable incomes are, beyond the bare necessities of life, more likely to be fun factors, which are simply not conveyed by contact with nature, whereas sport and entertainment are more likely to be.

Table 12.1 below provides a clear overview of the key actors and groups of actors.

Notes on Table 12.1: All these levels are meant when "nature conservation" is referred to in this book. Official nature conservation in particular is best captured through the legal exactness of laws and regulations. Where certain other actors are particularly central, they are named and highlighted accordingly.

However, it should be noted that "nature conservation" is not just an addition of social institutions or actors, but that it describes above all an attitude of mind. This book is thematically dedicated to this overall attitude.

Table 12.1 Actor levels of nature conservation

Players and their characteristics	Representative/ examples	Important measures and tools
Official nature conservation	Conservation officer Lower and upper nature conservation authorities at county level or in regional councils Federal Agency for Nature Conservation Relevant ministries European Commission	Laws, decrees Budget Restrictions, Penalties & Restrictions "Scientific" advice

Table 12.1 (continued)

Players and their characteristics	Representative/ examples	Important measures and tools
Education and research institutions	Professorships and chairs at (technical) colleges and universities; but also: Federal Agency for Nature Conservation (BfN), relevant state research institutes	Researches, establishes norms and discusses values Conservation History "Conservation professionalism" Training of experts
NGOs (non-profit, non-governmental organisations)	NABU, FOE, WWF	Mobilization of the public Contact with the mass media, donations and memberships
Public and private foundations	German Federal Environmental Foundation (DBU), Heinz-Sielmann-Foundation	Purchase and maintenance of areas for nature conservation purposes, financing of nature conservation projects
Political parties	The Greens	Voters, political bodies, influence on legislation
Individual representatives of nature conservation, especially those interested in nature and unaffiliated committed people	(Individuals) Often (but: not always) introverted and escapist character	(As little networked and numerically negligible individuals of low social influence)

13

Nature that Does not Deserve Protection: Spontaneous Vegetation, Ruderal Communities, Neophytes and Neozoa

An Inventory

In Neuenheimer Feld in the north of Heidelberg, excavation is underway. The excavator leaves behind an initially barren brown clay surface that could also be the surface of an inanimate planet. It is autumn, and until spring nothing happens on the barren surface. Then, however, the area begins to green up, first sporadically, then increasingly over a wide area (Fig. 13.1). Among the spontaneous colonizers are many former field weeds, such as chamomile, cornflower and corn poppy. **New colonizers** continue to arrive until the end of summer. By the end of the year, some 70 plant species may be growing in the small area of about one acre. Even more numerous are the arthropods, mainly insects, that eat the leaves of these plants, pollinate their flowers, or hunt for other small animals. If the area has not yet been built over the following year, which is unlikely to be the case, a few more species will be added. At the same time, the vegetation

K.-D. Hupke, *Nature Conservation*,
https://doi.org/10.1007/978-3-662-66159-8_13

Fig. 13.1 Still patchy vegetation in the first year consisting of a few pioneer species on construction land. (Own recording)

begins to develop more upwards, in a dense tangle of mugwort and evening primrose, for example. There are also some bushes and tree pioneers like black locust, willow species or dogwood. In addition, there are plenty of blackberries. Individual species that are particularly in need of light are already disappearing again under the increasing density and shade of the vegetation. If the development were to be continued without further external intervention, a high-stemmed forest would develop within a few decades. Measured against the conditions in the first and second year of the fallow, this would probably be poorer in flowering plants overall, because the tree layer, which is always naturally poor in species in Central Europe, would prevent a stronger development of the herbs close to the ground.

The **initial biotic community** just described (cf. Figs. 13.1, 13.2 and 13.3) is about as species-rich as the southern German calcareous grasslands and calcareous fens that are particularly favoured by nature conservation (for comparison, see Table 13.1). And yet, surprisingly, such

Fig. 13.2 Ruderal vegetation along a former railway track. In this case, the species diversity is further enhanced by the different grain sizes of the ballast body, running zonally from bottom left to top right in the picture, and by their different site qualities. (Own recording)

Fig. 13.3 Colourful flowering aspect of a species-rich ruderal meadow in early summer—no object for nature conservation? (Own recording)

Table 13.1 Grasslands and ruderal areas—What are the differences?

Comparison point of view	Rough grassland (vs. [semi-] dry grassland**)**	Construction land
Stability	Stable over years, with cessation of grazing and mowing gradual emergence of scrub and eventually forest	unstable, rapid succession to forest in often only a few years
Physiognomy	Mainly herbs, less grasses, mostly low-growing, only a few taller perennials	From the beginning many taller herbs like mullein (*Verbascum* spec.)
Origin	As a rule, remnants of a previous use, often going back centuries	Forms spontaneously and at short notice on all surfaces that somehow become "free"

Table 13.1 (continued)

Comparison point of view	Rough grassland (vs. [semi-] dry grassland)	Construction land
Species constancy	Very high	Very low, many species are quickly replaced by new ones from one year to the next in the sense of succession
Need for protective measures	Absolutely, otherwise there would be a gradual transition to forest; the disappearance of this habitat would be irreparable even in the medium term	Despite rapid transition to forest when use ceases, no need for protection, as comparable areas can be created anywhere and at short notice, e.g. during construction work
Growth strategies of plants	Often very slow growing, advanced in succession (K-strategists)	Fast colonizing, spontaneous, fast growing, at the beginning of the succession series (R-strategists)
Ratio archaeophytes/ neophytes	Almost no neophytes (in intact rough grassland)	Neophytes very numerous and often dominant in terms of species and individuals
nutrient supply, especially of physiologically effective nitrogen compounds	Low	Mostly high
Spatial distribution	Often in the marginal areas of the communal land, in this case succession of old Allmend pastures; preferred on slopes difficult for agriculture or with shallow soils	In the middle of settlement areas, often on new construction sites, or linearly along roads and railway lines

biotic communities are neither the subject nor the concern of nature conservation. This is due, on the one hand, to their clearly human-induced, "unnatural" character. Who wants to (and is able to) protect construction land? Furthermore, such communities are not old; and for nature conservation everything that is old is somehow valuable. On the other hand, this biocoenosis obviously does not need special protection, because such **fallow and ruderal areas** (from the Latin rudus = rubble) are constantly being created within and on the edges of settlements anyway. Also, in order to protect this species richness, one would have to constantly create new initial areas in the form of dredged areas, for many nature conservationists an absurd process.

Many of the plant species found in "disturbed" ruderal areas are also to be regarded as **neophytes.** Nature conservation has always been sceptical about these new species, which have so obviously immigrated under the influence of humans and thus disturb the illusion of having an "untouched" nature in front of them. However, almost all neophytes settle primarily on ruderal areas or other young succession stages, which are little appreciated from a nature conservation point of view, and thus pose little competition at least to the inhabitants of calcareous grasslands and calcareous fens (the native communities richest in flowering plant species). Neophytes almost always colonise sites that have already been "disturbed" by humans in the sense of eutrophication (excessive nutrient enrichment) or damage to the local vegetation. The neophytes that settle there correspond in many respects to the ecological conditions of this new site and, ecologically speaking, belong exactly there (otherwise they would not exist there). And they clearly expand the species range of the native flora (scientific observations in this respect in recent times, for example, in Hejda and de Bello 2013, pp. 890–897). That conservationists generally do not value this expansion is because, for them, all species are precisely not equal (Chap. 10). The conservationist values. This clearly distinguishes him from the scientific approach, which observes and explains, but by method and by definition closes itself to value judgments and sense judgments. This circumstance illustrates the difference between natural science and nature conservation.

This circumstance is particularly remarkable because the species that immigrated before 1500 **(archaeophytes)** are often virtually preferred in nature conservation. This is also due to the fact that they are dependent on open, nutrient-poor sites, such as those that resulted from extensive farming practices within traditional agriculture. However, it is precisely these areas that are particularly endangered by intensification of agriculture and afforestation, and with them their large number of species of flowering plants. One could also say pointedly: archaeophytes are not only tolerated in nature conservation, they even form the main content at least of traditional nature conservation ambitions.

What is striking is the particular commitment with which authorities and politicians pursue the suppression and **eradication of neophytes,** even outside the realm **of** official nature conservation. It seems to be out of all proportion to the apparent dangers to biodiversity. After all, hardly any neophyte can be shown to have succeeded in spreading into species-rich native communities; almost all are confined to "disturbed" sites of young successional stages, which—as already pointed out—are not target units of nature conservation anyway. The choice of words used in this "struggle" is reminiscent of war; for example, invasion is often mentioned, or the advance of the alien species must be stopped. Some species, such as the Caucasian Giant Hogweed (*Heracleum mantegazzianum*; Fig. 13.4) or the Mugwort *Ambrosia (Ambrosia artemisiifolia)*, are particularly emphasized as dangerous to humans in order to lend

Fig. 13.4 Fruiting stands of giant hogweed *(Heracleum mantegazzianum)* next to a public bicycle parking area. The plant already seems to extend one arm of its umbel threateningly towards the bicycles. The author does not dare to judge here how much of the threat scenario by the "invaders" is really close to reality. It is striking that many neophytes are assumed to be dangerous in principle: In this case, when the sap of the plant is pulverized, a sensitization of human skin to ultraviolet solar radiation, which could more easily lead to sunburns and thus skin damage. (Own recording)

Caucasian Hogweed

Giant hogweed can even be dangerous to humans. Since it was introduced at the beginning of the twentieth century, it has spread along riverbanks and forest edges, displacing native plants. "The sap of the plant is phototoxic and can cause burn-like injuries when it touches the skin in sunlight," explains Jan Thiele, a landscape ecologist at the University of Münster. "Giant hogweed, also known as Hercules' weed, has long been found everywhere in our country and can form dense stands that shade out and displace low-growing species." Mugwort ambrosia is similarly problematic. Introduced from North America in 1860, it has spread very rapidly to northern Germany in recent years. "Its pollen has a particularly strong effect on allergy sufferers," says Thiele (Deutsche Bahn 2012, p. 80).

emphasis to eradication (see box Caucasian Giant Hogweed).

While (thank goodness) **xenophobia** towards foreign immigrants is considered politically incorrect or at least mostly problematised in the established parties, xenophobia in nature conservation can develop largely unhindered. The parallel to human xenophobia is already astonishing on the semantic level: foreign things spread uncontrolled (!), they are more prolific than the native creatures, they are a danger to them. The suspicion is that xenophobia has created an outlet for itself here.

To repeat: scientific ecology as part of the biological sciences does not distinguish between native and alien species in the sense of a hierarchy of values. Values and norms are set (inter-)subjectively and are not part of the scientific discourse.

The classic goal of nature conservation is to maintain the **ecological status quo** while underestimating the dynamics of ecosystems that already exist “by nature”, with the consequence that man must continuously help nature to maintain its ideal state intended by man.

Two examples of **neozoa** escaped from captivity are shown in Figs. 13.5 and 13.6. Birds are particularly favoured for escape and resettlement because of their mobility. Both species have formed large and stable populations that show a tendency to continue to spread. Another example is the muskrat (Fig. 13.7).

Fig. 13.5 The Lesser Alexandrine *Parakeet (Psittacula krameri)* originates from the Old World tropics. In recent decades, however, it has spread as a captive escapee to regions of Europe and North America with mild winters, and today it is impossible to imagine the cities of the northern Upper Rhine region without it. (Own recording)

Fig. 13.6 The Egyptian goose *(Alopochen aegyptiacus),* native to Africa, has spread massively in parks and riverside meadows in recent decades. (Own recording)

Fig. 13.7 A muskrat as a neozoe on a riverbank fortified by stone packs. The approximately guinea pig-sized rodent from North America has only become native to Central Europe since the beginning of the twentieth century. For a long time it was discussed as a problem (destruction of reed stands, undermining of dams by earthworks). (Own recording).

Economic Loss

Non-native species cause significant economic damage to agriculture, forestry and livestock production, resulting in famine. Many invasive species have negative impacts on human infrastructure and affect human social well-being. Invasive species can reduce the appearance, accessibility and recreational value of a habitat (Nentwig 2010, p. 73).

Author's note: This should be contrasted with the positive economic effects of neophytes. For example, the robinia (also: false acacia) as well as the Indian touch-me-not *(Impatiens glandulifera)* have gained great importance in Central Europe as a source of nectar for beekeeping ("acacia honey").

The Fight Against IS in the EU

The danger is so great that the Brussels Commission is already considering *tighter border controls,* even between member states. About 12,000 plant and animal species are in the process of *conquering* Europe. On Dec. 13, 2013, EU environment ministers addressed for the first time the *advance of alien* species, which in official EU parlance are called *"aliens."* These include the Asian giant hornet and the Asian tiger mosquito. Some of the plants—such as the knotweed *(Fallopia japonica),* the late goldenrod and the giant hogweed – are even to be banned *from* trade in Germany.

According to the EU Commission, it has long since ceased to be just a matter of suffocating biodiversity, but of causing tangible economic *damage*. The Japanese knotweed *damages* buildings, the signal crayfish is considered a carrier of crayfish plague and *destroys* native crayfish species. In Brussels, the *damage* to farmers, homeowners and social security funds is already estimated at €12 billion. Unlike Australia and New Zealand, where *laws against* the bio-invaders have been in place for some time, Europe is moving late. As the insects and plants do not stop at *borders*, joint action by the 28 EU states is needed, the ministers stressed. On the list of possible actions is a complete *ban* on import, sale, cultivation, breeding, use and release—at least for those species considered particularly *aggressive. In* addition, *border guards are* to carry out targeted *searches of* holidaymakers and business people. An EU-wide *surveillance system* is needed.

(*Niedersächsische Allgemeine*, 14.12.2013, quoted from Bundesamt für Naturschutz 2014, Natur und Landschaft 89, H. 2, p. 94).

Author's note: All words in the terminology of war and criminology have been italicized.

Fight Against the North American Grey Squirrel

Time is of the essence. The grey squirrel *(Sciurus carolinensis)* has not yet been seen in the wild in Central and Northern Europe. In northern Italy, however, it has already established itself on a broad front after an Italian diplomat released two pairs introduced from the USA in the park of his villa near Turin in 1948—where the animals did not stay, of course. Many populations are now established, such as in Piedmont and along the Ticino River near Milan. "The grey squirrel is currently spreading in Lombardy but has not yet reached Switzerland," says Sandro Bertolino of the University of Turin. Nevertheless, the Swiss authorities are "very concerned", especially as a grey squirrel population lives between the southern branches of Lake Como, just 15 km from the border at Chiasso. "Since the squirrels spread along forests or rows of trees, we assume that they still have 20–25 km to cover," notes the zoologist. It can't be long before they cross the border, because under favourable conditions the animals can extend their habitat by up to 250 km^2 each year.

For the squirrel, the further advance of the highly competitive competitor into northern Central Europe would, at worst, mean the end—at least it is threatened with an ever endangered niche existence. A new project of the European Union called EC-Square for the control of the grey rodent should now at least contain the danger. The aim is to eliminate or at least thin out populations of the grey squirrel that are easy to control so that they do not spread any further. To this end, animals caught in traps are euthanized on site with as little stress as possible (Walter Schmidt in Stuttgarter Zeitung, 13.3.2012).

14

Process Protection as an Alternative and a Silver Bullet?

Change of Scenery

A small piece of forest on a slope. Fallen trees lie criss-cross on top of each other. Piled up root plates tower meters high against the forest canopy. Wild alternation of giant trees, medium-growing trees and young growth. Lying tree corpses are overgrown with mushrooms, often also with young tree seedlings. It is difficult to make one's way through this wilderness. Some of the deadwood is so rotten that you break in as soon as you set foot on it. Other trees are still standing upright, but are rotted out hollow inside and are in various stages of collapse. Some have cancerous growths on the trunk. Old and young interpenetrate in a wild confusion.

This or something similar could be the description of a mature **ban forest** as designated in the natural forest cells/ban forest programmes of most state governments since the 1970s. But this is not yet the case. At least one generation of trees must pass before anything approaching a native

K.-D. Hupke, *Nature Conservation*,
https://doi.org/10.1007/978-3-662-66159-8_14

state will be restored. But even then, a static condition will not occur, such as in the sense of the obsolete ecological balance. **Windthrows** will tear clearings into this "primeval forest", light-flooded sites where **succession**, starting with tall shrubs and bushes, via one or two pioneer tree species, will finally enter a (temporarily) more stable climax state. In many places, however, no final stability is achieved: due to natural disturbances such as windthrow, insect calamities or, on slopes, landslides.

But why is this state of "wilderness", uninfluenced by humans and already aesthetically rejected by most foresters and forest owners as "unkempt" or "untidy", so much preferred by many conservationists—and they are becoming more and more so? After all, in the *National Strategy on Biological Diversity* in 2007, the German government decreed for itself to permanently exclude 2% of the country's land from use and leave it to its own devices by 2020 (Bundesministerium für Umwelt, Naturschutz, Bau und Reaktorsicherheit (BMUB), 4th ed. 2014, p. 31).

Here, too, the reason for creating **wilderness areas** lies in a high **species richness** (for forest ecosystems). But how does this come about in comparison with managed forest areas? The main reason is the fact that trees die a natural death here. Old wood, however, is an ideal nutrient substrate, e.g. for many fungi. On the other hand, the larvae of at least 1000 native beetle species live in **deadwood** (Figs. 14.1, 14.2 and 14.3), including those of the largest Central European beetle, the stag beetle *(Lucanus cervus).* Other target species highly valued in nature conservation also live on dead wood: for example, the rhinoceros beetle *(Oryctes nasicornis)* and the hermit *(Osmoderma eremita),* which has become known throughout Germany under the name Juchtenkäfer since Stuttgart 21 and almost prevented the largest railway project in Central Europe. All these species have become rare in **commercial forests**; as rare as trees

Fig. 14.1 Deadwood does not always remain upright for long, as in this picture. Previously perceived as unaesthetic under the guiding principle of the commercial forest, it now appears more as a decorative element. The knowledge of its "ecological" significance also has an aestheticizing effect. (Own recording)

Fig. 14.2 The larvae of the stag beetle *(Lucanus cervus)*, the largest Central European beetle species, also depend on deadwood. They live, among other creatures, in the stumps of felled strong deciduous trees, which in commercial forests require very late and thus often unprofitable "harvesting dates" of the wood. (Own recording)

Fig. 14.3 A larger-than-life artistic installation at the former State Garden Show site in Bietigheim-Bissingen illustrates the importance of deadwood for the beetle fauna. (Own recording)

that die of natural old age. Such old age death is completely contrary to the self-image of forestry, since every tree should be harvested first. A clash of interests between forest managers and nature conservationists (whereby the former often pretend to be the latter) can be demonstrated by this example.

The protection of natural forest development is also frequently apostrophized as **process conservation** (Scherzinger 1996). The aim of this form of nature conservation is, in contrast to the still dominant use of management to protect the status quo, to simply leave nature to its own devices.

Not that natural forests are close to nature from the outset. But they are on their way there. Under these circumstances, **closeness to nature** refers less to specific conditions that have been achieved than to a natural **process of succession that is** not controlled by humans. This is precisely what is worth protecting. Whether, on the other hand, forests protected in this way will ever return to their "original state" is certainly questionable in individual cases. After all, man has changed the location thoroughly in most cases, if one thinks of the soil profile, for example. In addition, the mostly small natural forest areas are embedded in land use structures with nitrogen inputs or a game density that is often too high for natural forest regeneration.

This contrasts with international efforts by nature conservation in the wake of the **Conference on Environment and Development in Rio** (1992) to strive for the **sustainable use of** forests as a conservation goal. Activities in this direction are aimed at **certifying** the use of tropical forests and forests in temperate latitudes. Ideally, every piece of wood can be traced back to the individual trunk and identified as sustainably harvested.

Both the sustainable use of forest areas and the designation of unused forest areas as forest protection strategies have their argumentative light and dark sides. Assuming a given demand for wood and wood products, a high proportion of unused forest means that the pressure of use on the remaining forest areas is likely to increase. Admittedly, the question arises pointedly as to whether we want to have near-natural commercial forests or unutilized forests next to timber plantations on a roughly comparable area of land.

Rare Mushroom Discovered in the National Park

Mushroom researchers have for the first time discovered a rare mushroom (Antrodiella citrinella) in the state of Baden-Württemberg. The mushroom grows in the Bannwald forest at Wilden See (Black Forest National Park). Experts regard the discovery as a minor sensation. The fungus only grows on another fungus, which in turn only grows on dead spruce wood. Due to these complicated growth conditions, the lemon yellow meadow mushroom is extremely rare. Until 2010, it was only found in Germany in two small primeval forest reserves in the Bohemian-Bavarian border region. Then the fungus appeared in the Bavarian Forest National Park, where it is now more common than anywhere else in Europe. The Mushroom is considered to be a "nature indicator", i.e. it only grows where nature is allowed to be nature and a certain nativeness has developed. This is certainly the case in the Bannwald at Wilden See, where nature has been left to its own devices for more than a hundred years.

(Bundesamt für Naturschutz 2015; Hg.: Natur und Landschaft H. 1, S. 39 [according to: Südwestpresse v. 23.10.2014])

Current Data on Natural Forest Development in Germany

In Germany, there are currently 213,145 ha of permanently protected forest with natural forest development. This corresponds to a share of 1.9% of the forest area in Germany. By 2020, the share is expected to rise to 2.3% and thereafter to approx. 3%. These are the results of a current research and development project of the BfN [Bundesamt für Naturschutz; d.Vf.] on the subject of "Natural forest development as a goal of the National Strategy on Biological Diversity".

The National Biodiversity Strategy (NBS) targets natural forest development on 5% of the total forest area or 10% of public forests by 2020. [...].

Forests with natural development comprise forest stands without direct human influence. The permanent cessation of forestry use and the refraining from interventions to secure nature conservation objectives on a delimited area of at least 0.3 ha in size are basic prerequisites for this. Forests with natural development also include those forested sites where all human use has ceased permanently and forest succession is foreseeable (Bundesamt für Naturschutz 2013b, p. 522).

In principle, the well-educated, older people and women are closer to nature and more sensitive to nature conservation than the formally simply educated, younger people and men. In the case of wilderness, the picture is different: Here, in addition to high school graduates, it is younger people and men who are particularly sympathetic to wilderness (Bundesministerium für Umwelt, Naturschutz, Bau und Reaktorsicherheit und Bundesamt für Naturschutz 2014, S. 12) (Table 14.1).

Table 14.1 Intensity of nature conservation interventions

Degrees of intervention	Related to habita t/biotope	Related to individual species
I: "Cheese bell" nature conservation	"Conservation by omission": protection of primary landscapes or process protection/succession protection	Don't hunt, don't gather, don't disturb
II: Stabilization; Indirect human influence	"Maintenance" by pruning, by mowing or grazing: protection of traditional agricultural landscapes: Stabilization of the status quo of succession	Installation of nesting aids etc.

(continued)

Table 14.1 (continued)

Degrees of intervention	Related to habita t/biotope	Related to individual species
III: Active transformation; direct human influence	Excavation of the uppermost soil layer or spreading of non-native soil substrates; "reset" of succession	Breeding in botanical gardens and zoos, with and without subsequent release into the wild; Seed and genome banks

15

Nature Where no One Expects it: In the City

Years ago, the Munich ornithologist Josef Reichholf attracted the attention of an interested public with his statement that within Germany, the city of Berlin had the greatest density of bird species; within Bavaria, this was the case with Munich (Reichholf 2007, p. 17). In comparison, the intensive agricultural landscapes of Lower Bavaria or northern Upper Swabia, for example, have only relatively low species densities. Similar results were obtained from species counts of flowering plants (Sebald et al. 1993, p. 10) for Baden-Württemberg. The northern Black Forest and northern Upper Swabia have lower species numbers on the same area than the highly dense central Neckar region, even if geological differences also play a role here.

However, a differentiated view must be taken of **species density in cities.** On large areas that are asphalted, concreted or covered with roofing, the species density is close to zero. The density of naturally occurring (i.e. not planted) species is also low on many small green spaces and in front

K.-D. Hupke, *Nature Conservation*,
https://doi.org/10.1007/978-3-662-66159-8_15

Fig. 15.1 A pre-existing woodland has become established within the commercial brownfield site. (Own recording)

gardens. By contrast, both zoological and botanical diversity is high in old cemeteries, in the siltation zones of often man-made bodies of water, in old, often unkempt parks, in railway and industrial wastelands (Figs. 15.1, 15.5 and 15.6), and in forests managed close to nature for aesthetic and recreational reasons, which often surround large cities like a ring. The fact that there is **no pressure** from agriculture and forestry in the vicinity of large cities has a particularly positive effect on biodiversity. However, as has already been shown, this is the main cause of the decline in species in the modern cultural landscape.

This diversity can be demonstrated by the example of **the** overall manageable **number of species of larger mammals.** Today, wild boar *(Sus scrofa)* and red fox *(Vulpes vulpes)* can be found even in inner-city areas, both also often in allotment gardens, where almost no outsider would have suspected them. Wild boars pass through here on nightly forays and can cause great damage. Foxes, on the other hand, are often hardly noticed and often have their den in the shelter of a shrub thicket or a hedge. In many cases, hares *(Lepus europaeus), which are* exposed to a strong decline in populations in the fields, also hop across green spaces or, as in Heidelberg, across the university campus.

Among the birds, the relatively large birds of prey are conspicuous. The *peregrine* falcon *(Falco peregrinus),* formerly considered a cultural fugitive, now regularly breeds on a church tower in the inner cities of Esslingen or Heidelberg, for example. Buzzards *(Buteo buteo)* do not feed on mice in the cities, but on animals run over in traffic or on waste.

However, the diversity of plants in inner-city areas has also increased over the past decades. This is mainly due to the fact that the demands on the care of plants have decreased in many respects. For example, many relatives no longer care much about the upkeep of graves. With life partners working double jobs, there is often no time for garden maintenance. Property managers have discovered that an unfertilized lawn requires less frequent mowing, thus saving money. In the latter case, however, as nutrient salts are gradually lost through leaching, the way is virtually paved for greater biodiversity. In a suburb of Stuttgart, for example, I was able to find a total of three species of orchids (Orchidaceae) growing wild on lawns in a single domestic garden (Fig. 15.2). The only counterproductive factor here is the heavy input of dog excrement, especially along roads and paths, which leads to overfertilization (eutrophication) of the sites.

In particular, our ecological-aesthetic ideal of an **inner-city green space,** be it a public green space, a private garden or an old cemetery, will decide how much biodiversity we will allow. In principle, a clearing of dense tree growth or shrubbery is at least rather beneficial for the biodiversity of flowering plants, and thus also of many insects; but only if at the same time there is a lack of nutritive salts as well as a tolerance towards "weeds" or all non-lawn plants, which today are often more aptly labelled as wild weeds.

Fig. 15.2 Orchids growing wild in a front garden (bee ragwort, *Ophrys apifera*). (Own recording)

The new image of the city as a **focal** point **of biodiversity** is difficult to reconcile with the traditional thought patterns of nature conservation, which, as has already been explained in detail, are primarily oriented towards forested areas and traditional, extensively used agricultural landscapes. Of course, the protection of a striking individual tree or an old cemetery avenue as a natural monument is also provided for now and then in this concept—but the city as a whole as a target area for nature conservation? The dominance of economic and social zoning seems far too clear for that, the natural elements appear far too isolated in view of the abundance of built-up space. And yet cities will (have to) be one of the focal points of future nature conservation, especially since there is an increasing lack of nature conservation alternatives outside the settlement area. In addition to the protection of animal individuals, such as the peregrine falcon nesting on an urban church tower, it is also a question of considering how even more nature can be drawn into the city, for example by ensuring that the urban development structure takes this into account. And the prospects for this are not bad: after the Modernists (Gropius, van der Rohe) initially engaged in an internal dispute as to whether trees were at all sensible or permissible in the city, a younger generation of urban planners has now become very open to **nature in the city.**

And far more could be done without great additional cost or land requirements if instead of laurel cherry *(Prunus laurocerasus)* and conifers, more deciduous woody plants were planted in parks and gardens (these may well be exotic ones, such as the Japanese ornamental quince (*Chaenomeles japonica),* and lawn clippings were consistently removed. This would lead to a thinning of the lawns and an increased species richness of herbs, which in turn would bring about a diversity of insects such as wild bees, beetles and butterflies (Fig. 15.2). The popular garden ponds (Fig. 15.3) can

Fig. 15.3 This simply constructed garden pond in a plastic tub of about 150 L is framed by boulders weighing about 30 kg each, which gives a natural impression. However, in this case, the boulders also provide refugia for the pond newt *(Lissotriton vulgaris)* over the winter. This small newt species, which prefers open, sunny small water bodies and which was common only a few decades ago, has declined considerably due to the mass reproduction of grey herons, and in some areas also of storks. Here in the settlement area of an allotment garden the grey heron occasionally appears, but nowhere near as often as outside. So the newt still has a chance. (Own recording)

also contribute a great deal to biodiversity in settlements if they are designed in a near-natural way.

However, it is particularly important to refrain from fertilizing green areas (Figs. 15.4 and 15.5).

Fig. 15.4 Fertilization of a public green space: counterproductive for biodiversity. (Own recording)

Fig. 15.5 Not a Tibetan pilgrimage trail, but a deterrent measure against the little ringed plover, which likes to choose fallow land as a breeding site. If a brood of the endangered bird is detected on an area, it may not be built on under nature conservation law until the brood has fledged; an expensive matter for the investor. (Own recording)

Christmas Market in the Castle is Cancelled

Bad news for all fans of the Heidelberg Christmas market at the castle: It will no longer take place this year and probably also in the coming years. The main reason for this is the bats, which traditionally hibernate in large numbers in the protected corridors and walls of Heidelberg Castle. Five years ago, a large Christmas market was held for the first time. Since then, it has attracted more and more visitors each year—most recently around 60,000.

According to an expert opinion of the Regierungspräsidium, the populations of bats in the area of the Stückgarten, where the market took place, have decreased by 50% since its introduction in 2010. As nature conservation law prohibits disturbance of the bats' hibernation, the cancellation of the event was apparently unavoidable. The expert opinion had shown that the bats in their winter quarters had suffered from the Advent restlessness and that their existence could be seriously endangered in the medium term as a result, reported the managing directors of the State Administration of Palaces and Gardens Michael Hörrmann and Andreas Falz at a press conference in Heidelberg. Therefore, the market will be suspended for the time being until more facts are available.

The decision was painful, but "necessary and unavoidable", Hörrmann said. It had been clear from the beginning that the market was only possible with a constant eye on species protection, Falz said. "The bat population of Heidelberg Castle is unique in Europe and must be protected," he said.

(Stuttgarter Zeitung v. 20.10.2016).

Public Green Spaces as a Horticultural Experimental Field with Natural Elements

Only a few decades ago the distinction was clear: a green space consisted of a well-kept lawn, in which not much was tolerated that did not belong to the respective sown graminaceous plant. In between there were flowerbeds, say of dahlias and gillyflowers, of roses and tagetes, whose unnatural character was also beyond doubt. Between these no "weeds" were allowed to grow at all. For nature conservation these were almost worthless areas.

The concept of the horticulturally valuable has changed, however. Today's garden concepts no longer see a garden as an antithesis to wildly rampant nature, but rather want to give it a certain amount of freedom, within limits, of course, which ensure that the gardener does not become superfluous. He is needed anyway, if only to protect buildings and infrastructure from the sprawling greenery that constantly has to be contained and cut back to its intended state: to protect paths, rails, roads and facades.

Plants immigrating into parks and green spaces are now quite tolerated within this new horticultural self-image: the former weeds are upgraded to wild herbs.

In many cases, however, this is not enough for public gardeners. Wild herbs are deliberately sown; because of the greater amount of work involved, they are planted less frequently. Some of them would sooner or later establish themselves on this site anyway. Many are virtually indistinguishable from the wild forms native to the region. Others have significantly enlarged flowers or flowers with an unusual hue and are obviously forms bred specifically for that purpose. Overall, it is mostly a wild mix that promises "more nature".

This new horticultural aspect is difficult to assess for conservationists, who must always be thought of as valuers and doers. In the catalogues of the "wildflower" seed dealers there are also protected species again and again. These may certainly be propagated and resold in a commercial horticultural enterprise, provided that the initial seed has been acquired in a legally impeccable manner. Also a breeding transformation is not forbidden. When sown "in the wild", the cultivated forms mix with any wild forms that may be present in the region. If the forms of the sown cultures do not differ too much from the wild form, they are now again taken into custody by the Nature Conservation Act. Often this also depends on the region: An edelweiss *(Leontopodium nivale) that has been* released into the wild in northern Germany will certainly not fall under the Nature Conservation Act; in the Upper Bavarian foothills of the Alps, however, it will.

Botanists and conservationists alike are generally opposed to such horticultural "beautification" or "enhancement" of nature. Of course, they are aware that nature in Central Europe is always influenced by humans. Nevertheless, value is placed on a natural self-dynamic, as expressed for example in the conquest of new sites, and not on horticultural planting or sowing.

The anointing of new species has also become very important for pond management. Mail-order companies and garden centres offer a wide range of marsh and aquatic plants, most of which have not been further developed through breeding, i.e. are directly derived from specimens taken from the wild. However, these regularly come from populations that are not characteristic of the region in question. They mix with native forms and thus have a decisive influence on the aquatic flora. A large proportion of the aquatic vegetation in southern Germany, such as calamus (a neophyte since the sixteenth century), ragged buttercup, swan flower, arrowhead, sea canary and crayfish claw, is derived from such deliberate salvaging in recent decades. An overview of the near-natural distribution of these species, which were often rather rare in southern Germany only about 50 years ago, is thus hardly possible today.

In connection with the anointing of natural elements, the already difficult boundaries between nature and garden, between nature and culture, finally fade, which would make the conservationist as an advocate of nature in the completed case really superfluous. The original conservationist, who benevolently accompanies and, if necessary, protects natural dynamics, always under human influence, tends to become the gardener who, out of his own ideal, determines what has to grow where (Fig. 15.6).

Fig. 15.6 A railway wasteland has developed over decades into a diverse inner-city green space. Now this is to make way for the new location of the transport association depot. Many residents are upset; some have founded a citizens' initiative to prevent the project, from which the pictured poster originates, which was hung on a construction fence. Nature in the city is on the one hand a success story, insofar as more and more species originally foreign to the city are immigrating from the surrounding agricultural and forest areas. On the other hand, however, this increasing importance of urban nature is also threatened by redensification, as in the present example. (Own recording)

16

Military Terrain: A Natural Idyll?

Tanks roar, digging deep into the sandy-loamy soil with their tracks, tearing open new trenches again and again, ploughing through puddles, leaving behind loamy-yellow water that closes again behind the tracks. An abomination for every nature and landscape lover, the peak of nature destruction.

If you look into the loamy puddles, you will see the heads of yellow-bellied toads coming to the surface soon after the tank has passed through. These animals seem as impervious to tank tracks as they are to tractor tires. Presumably this is simply because the toads, like most of the water and puddle mud, are displaced by the chains to the side, where the toads survive the passage of the tanks through their biotopes quite unscathed—usually at least.

But the ruts of the tanks are certainly not an idyll overall. Only a few species have survived here, often specialists rare in nature outside, which do not tolerate competition from other species or shading by vegetation. The situation is

K.-D. Hupke, *Nature Conservation*,
https://doi.org/10.1007/978-3-662-66159-8_16

different outside the relatively limited lanes. Here, the military training requirement has kept a traditional agricultural landscape alive, which has not followed the development outside in the direction of increasing yields and intensification. Here, while the landscape has been kept open by preventing natural reforestation, a large variety of plant communities and species, as well as insects, amphibians, reptiles and birds, have survived and re-established themselves. Thus, despite continued **military use,** in terms of maintenance and structure, a close proximity to nature reserves can be observed, whose main focus is also on **open areas** such as rough pastures and rough grasslands. Here, too, it can be stated that, at least in our latitudes, **shading** and **nutrient richness are** the main enemies of a species-rich landscape.

There are certainly many objections to the sense and ethics of military training grounds. For nature conservation, these areas, although not specifically labelled as nature reserves because they are or were used for military purposes, are usually a real stroke of luck.

Therefore it is not surprising that the first wolf populations *(Canis lupus),* which migrated from Poland to East Germany, found shelter mainly in still existing or in former **military training areas.** The wolves have long since become accustomed to the noise of shooting exercises and the presence of soldiers, since they are not hunted.

Against this background, the numerous abandoned military training areas, especially in eastern Germany, have become an opportunity for nature conservation. Now nature conservation is faced with the task of imitating military use through maintenance measures. In western Germany, the Swabian Alb Biosphere Reserve around the former military training area of Münsingen can be considered an example of such a **conversion.** The small-scale alternation of wooded or tree-covered and open areas is particularly attractive in terms of landscape aesthetics and ecology.

In connection with the conversion of former military areas to nature conservation, the **"Green Belt"** (Fig. 16.1) is the successor to the "Iron Curtain" in the form of a band-like area several dozen to several hundred metres wide along the former border. In this area, which was formerly not used for agriculture, was not fertilised and was kept free of trees for reasons of visibility, numerous plant communities requiring light were able to survive. In addition to the biological richness, the former borderline also offered the opportunity of a coherent structure in the sense of linearly aligned and networked nature conservation areas (cf. Chap. 4), which should enable migrations of plants and animals.

Fig. 16.1 The former border between the FRG and the GDR as a "Green Belt." (© K. Leidorf; source: https://www.erlebnisgruenes-band.de/uploads/pics/GruenesBand_Mackenrode_ausschnitt_01.jpg)

Paradise Created by Tanks

About 20 years ago, the US Army training area in Böblingen's city forest looked like a lunar landscape: a bare area with deep craters gaping in the ground—torn open by the chains of the tanks that constantly crisscrossed the training area. Grass and bushes therefore had little chance to thrive.

What looked like a huge violation of nature to ordinary citizens, however, brought tears of joy to the eyes of conservationists. "A true gem of biodiversity has been created," says Rolf Gastel of the Nature and Biodiversity Conservation Union (Nabu). Yellow-bellied toads and natterjack toads settled in the small pools of water that formed in the tank furrows. On the bare, sandy areas, the dune sand beetle spread. All species that have all but disappeared elsewhere. The conditions on the training grounds are a paradise for endangered species.

[...]

Despite the practice operation or precisely because of this use—depending on your point of view—the site has become a refuge for rare animals and plants. "Many species cope better with the exercise operation than with conventional agricultural use," says Rolf Gastel. After the withdrawal of the tanks in 1994, he was one of the founders of a working group that wants to preserve these species. The problem is that without the tracked tanks, the area is rapidly becoming overgrown—and the yellow-bellied toad and dune sand beetle are losing their habitat.

Since then, voluntary nature conservationists, the US Army's Stuttgart Environmental Office, the Regional Council and the Federal Forestry Office have been working together to keep the heath-like areas of the site open. A shepherd moves with his animals over the 600 ha area. Large tracked vehicles regularly drive over the area and pull furrows into the ground so that further ponds can be created there. The caterpillars and tractors aren't quite ideal for maintenance, though, says Ina Gebhard of the environmental department at the U.S. garrison in Stuttgart. "Actually, we would need a maintenance tank to drive over the grounds" (Gerlinde Wicker-Naber, *Stuttgarter Zeitung*, 12.11.2012).

17

Second-Hand Nature: Renaturation of Quarries and Opencast Mines

Almost no human intervention has such a devastating impact on nature and landscape as the large **opencast mines** in Central Europe, especially in connection with lignite mining on the edge of the North German Plain towards the low mountain ranges (Ville, Lausitz). Areas the size of counties are deprived of their fields, forests and settlements. After a few years or decades the raw material deposits are exhausted; a barren "moon landscape" now stretches as far as the eye can see.

But the energy companies are legally obliged to **recultivate**, which can also be a **renaturation**. Previously excavated "topsoil" that was temporarily stored during the mining phase is reapplied. A recultivation plan then provides for either new agricultural land, forests or recreational areas. The latter are often located on water surfaces of the old open-cast pits that are gradually created by groundwater rise. And all types of land use can in principle be linked to nature conservation requirements, although areas with nature conservation status only account for a small percentage

K.-D. Hupke, *Nature Conservation*,
https://doi.org/10.1007/978-3-662-66159-8_17

of the total area. Nature conservation has a hard time in such an artificially created zone. At least in the first decades, only ruderal flora (Chap. 13) can be expected, which nature conservation is known not to attach much importance to. But there are exceptions. These are mostly located in the shore areas (later siltation zones) of the developing lakes.

Outside the area glacially overformed during the Würm Ice Age (= soft-sea ice age), Central Europe (like most regions of the earth) is by nature extremely poor in freshwater areas. Where these have been formed by natural factors such as glacial overforming, but also tectonic subsidence or by cut-off meander bends of river courses, they are, seen in geological time scales, extremely short-lived. Depending on depth, nutrient salinity and biological-climatic zone, these **inland lakes** have a life span of only a few hundred to a few thousand years. Abiotic sedimentation of clays, silts and sands as well as biotic sedimentation with peat and sludge ensure that the depression is gradually filled and the level of the surrounding area is adjusted. The originally existing still water has disappeared at the end of this process.

However, for centuries and with the growing technical possibilities, man has increasingly provided for new standing waters. This is done in a technically relatively simple way by damming up a flowing watercourse. Such **reservoirs** (or impounding ponds) have the disadvantage of an often strongly fluctuating water level, because man uses the impounded water in seasonally varying degrees for reasons of energy production or flood protection (in dry areas also because of irrigation). Such **water level fluctuations** over the course of a few weeks or months make it difficult for animal and plant species to zone along the lakeshore according to water depth. However, such **zonation** forms the basis for the overall species richness of the water body at natural water bodies with only minor water level fluctuations, as each of these zones develops its own characteristic vegetation and fauna. If the water level fluctuates strongly, as in the typical case, reservoirs are mostly poor in plant and animal species as well as individuals, which is particularly evident in the bare bank areas.

In comparison, **dredged lakes**, as they arise from gravel pits, quarries and opencast mines, are mostly characterized by a low seasonal fluctuation in the water level. This is due to the fact that here river water is not used by impoundment, but the water level is oriented to the **groundwater level of** the surrounding rock and sediment bodies. Dredging lakes do not need to be impounded, but fill up gradually, often only after years in the case of larger depressions, as a result of groundwater penetrating laterally up to their groundwater-dependent maximum level.

Lowering the water level in quarry lakes would only be possible by pumping out the water, which is technically complex and expensive. Dredged lakes are also spared from rapid and catastrophic draining by bursting of the dam. If the water level is stable, even a quarry pond that is free of vegetation in its initial stage can develop a great wealth of plant and animal species in a pronounced zoning, especially of the shallower shore zones, after only a few decades.

Initially almost vegetation-free areas at quarry ponds are an example of stable nature (Chap. 8, Fig. 8.1), which comparatively quickly recovers completely deserted areas. They are important for a considerable proportion of bird life (waterfowl and marsh birds), for numerous mammals (especially rodent species and bats), for several species of reptiles and almost all amphibians as well as almost innumerable arthropods (insects, crustaceans) and for numerous other taxa.

Using the example of **amphibians** that are particularly dependent on water, it can be shown that species such as the Great Pond Toad, the Yellow-bellied Toad and the Green Toad, which are particularly fond of sun and warm water, are found in Central Europe almost exclusively in such anthropogenic bodies of water (Figs. 17.1 and 17.2),

Fig. 17.1 Sunlit pond with little vegetation at the deepest point of an abandoned sand pit on clayey (waterlogged) subsoil. Habitat of the Yellow-bellied Toad. (Own recording)

Fig. 17.2 The Yellow-bellied Toad *(Bombina variegata)* is dependent on areas kept open by humans and prefers warmly sunlit waters with little vegetation. (© C. Wermter/blickwinkel/picture alliance)

and in some cases probably only migrated from the forest steppes of south-eastern Europe after they were created. It is precisely these species, which are threatened without human intervention, that are special target species of classical nature conservation.

Quarries and opencast mines are generally seen by nature conservationists as open wounds in the face of the landscape. Ecological compensation measures for transport projects in the Augsburg area in recent years show that this can also be seen differently. Here, the soil and the top layer of gravel were removed to a depth of about 1.5 m from former farmland. The gravel obtained in the process can be made available to the construction industry. The remaining shallow gravel bed is a succession area into which the rough grassland plants (Chap. 20, Section “Meagre Grasslands”) of the historically widespread Lechtal heaths are to migrate again in the long term. This is a long process. Initial successes are already evident. In other places, nature conservationists are vehemently resisting new gravel pits. Nature conservation, as in other contexts (cf. Chap. 7), is rich in contradictions.

18

Is Nature Only Intact if all Species Increase Equally?

In 2002, the **house sparrow** *(Passer domesticus)* was named "Bird of the Year" by the nature conservation organization NABU. The population of the bird, formerly considered a seed pest, had declined by more than two thirds. NABU wanted to draw attention to this **population decline as** well as to the fact that the house sparrow is threatening to disappear, especially from cities.

The actual home and original way of life of the house sparrow is not known today, as it has completely adapted to human settlements in its nesting and way of life. With man, the house sparrow has also spread to all continents, with the exception of Antarctica. It is extremely common in large parts of the world. Basically, it is still one of the most common bird species in Germany, but with a slight tendency to decline. The causes of this decline are, as so often with the always complex ecological situations, difficult to determine. One role could be played by the decline of dung heaps, which are traditionally often located between farms and

K.-D. Hupke, *Nature Conservation*,
https://doi.org/10.1007/978-3-662-66159-8_18

village thoroughfares. Especially on the dung heaps, which were warm even in winter due to fermentation processes, the sparrow could find worms and insect larvae in addition to seeds and other vegetable waste. This ensured its nutrition throughout the year. Today, dung heaps are largely a thing of the past due to more intensive livestock management and are generally considered unhygienic and unsightly. Today they are mostly replaced by parking lots or flowerbeds. On these, however, the sparrow finds little or no food.

At least the mass reproduction of the sparrow thus goes back decidedly on the human influence. It is also man who has now made the house sparrow rarer again in the wake of a structural-technical reorientation. It remains questionable why nature conservation declares one of the still most common native bird species to be an endangered species, which, incidentally, was also the reason for the sparrow's inclusion in the early warning stage of the Red List of Threatened Species.

The fact is that the house sparrow competes with other species for food and nesting sites. If one species becomes rarer, there is an opportunity for others. Since the house sparrow is still one of the most common species anyway, this should definitely be an opportunity for the diversity of the avifauna as a whole.

So what can we learn about nature conservation from the sparrow as "Bird of the Year"? That nature conservation only considers nature to be intact if now really all animal and plant species increase equally. One could also say that nature conservation in general does not understand ecology or does not want to understand it. At least if one understands it as a natural science according to a cause-and-effect concept.

Thus, if one follows its self-conception, nature conservation makes rare species more common, but common species should not become rarer in the process. Basically, nature conservation needs a growing planet.

On the other hand, one can also see how much nature conservation is oriented towards the ideal of "the way it was in our childhood" or "always the same". With these static guidelines, however, it does not do justice to modern scientific ecological approaches of succession and variability (Jax 1994).

The Blue Tit is the Winner

The blue tit is more common in the gardens of the southwest than last year. "Among the top 10 species, we only see a slight upward trend in the blue tit with plus twelve percent," said Hannes Huber from Nabu Baden-Württemberg on the balance of the current bird census. Species such as the sparrow (*Passer domesticus*), great tit (*Parus major)*, redstart (*Phoenicurus ochruros*) and magpie (*Pica pica*) are at about the same level as last year. The birds survived the occasionally very cold winter well. In the northern part of the Upper Rhine, however, the Usutu virus had greatly reduced the number of blackbirds (*Turdus merula*). Overall, however, the blackbird population is stable. During the "Garden Bird Hour" in mid-May, around 3700 Baden-Württemberg residents counted the birds on balconies, in gardens and parks and reported more than 78,000 observations to Nabu (Stuttgarter Zeitung, 01.06.2012).

Note: Emphasis in *italics* and bold by the author.

Humans are not always directly to blame for drastic declines in a species. Sometimes it is nature conservation itself. In the 1970s, for example, the ponds were richly stocked with newts. At the same time, the grey heron had become extinct or extirpated except for one breeding colony on the northern Upper Rhine. Storks had also disappeared except for small numbers, especially on the Upper Rhine.

Consistent protection of the grey heron has now made it quite common again since the late 1980s. Pond newts are hardly found any more. Green frogs, which prefer vegetation-rich riparian zones of larger water bodies, are somewhat better protected; but they too have declined significantly.

However, the decline of its main prey has by no means made Grey Heron populations rare again in terms of simple predator-prey relationships. Grey herons, like many other birds such as storks, crows and gulls, are quite opportunistic when it comes to feeding. They can be found at urban waste sites as well as in freshly ploughed fields.

Also in the 1970s, the walker was repeatedly frightened by the powerful wing beats of a troop of partridges in open fields. There were not many foxes, because for reasons of rabies control their burrows (and due to the lack of easy distinguishability of the inhabitants from the outside also those of the badger) were regularly fumigated. In the meantime, foxes can be found everywhere and are probably our most frequent predators: in forests, in open agricultural landscapes, in urban green spaces and parks, in allotment gardens and in old cemeteries. At the same time, the grey partridge has disappeared almost everywhere. As a ponderous animal with limited flight capabilities, it is a grateful prey of the fox. Like the grey heron, the fox is an opportunist in its diet, not simply becoming rarer as the number of its main prey decreases; it also goes after carrion and human waste.

The argument of an increasingly "cleared" agricultural landscape, which is often put forward for the partridge decline, is not convincing either. This process is already more than 150 years old, and has had little effect on the partridge population for more than 100 years. Moreover, field hedges have not declined to the same extent as partridges and are still widespread in high densities in some sub-areas.

Both examples show that nature conservation (and this seems particularly obvious in the first example) does not simply "protect nature", but also changes species frequencies and the numerical proportions between species; often to the detriment of some species that are certainly threatened and "worth preserving".

19

Nature Conservation is Quite Successful: The Example of Large Animal Species

Perhaps sometime in the third decade of this millennium you will take a spring walk in a floodplain along a stream or river. You may discover that in the middle of a designated nature reserve, a professional logger has systematically cut down a number of willows or poplars in a round cut. If you are alarmed and suspect that this is an unauthorised encroachment on a nature reserve, you can probably be reassured. It is probably just the work of beavers, our largest native rodent.

Originally, the **European beaver** *(Castor fiber)* was widespread along the floodplains from the lowlands into the low mountain ranges. However, because its fur was valuable and it was considered a forest pest, it was persecuted and eradicated almost everywhere in Germany until the nineteenth century. Against the resistance of many foresters, however, it has been reintroduced several times since the 1970s. From these initially few and selective settlements, the beaver has spread over a wide area, following the course of the watercourses. In the meantime, it can be considered firmly

K.-D. Hupke, *Nature Conservation*,
https://doi.org/10.1007/978-3-662-66159-8_19

naturalized and can once again be found in larger populations. However, its logging activity is such a nuisance that the animals, which are protected in themselves, occasionally have to be captured near settlements and released again in more remote places: a strategy that will certainly no longer be effective at some point if the beavers continue to multiply.

The example of the beaver is a particularly good illustration of how a species is closely intertwined ecologically with the living requirements of other species. More than almost any other animal species in Central Europe, the beaver creates its habitat through its own activities. With its sharp incisors, it cuts down softwood trees near streams and builds a beaver dam from them, reinforced with fine twigs (Figs. 19.1 and 19.2), thus damming the stream. In the beaver pond thus created, it creates underwater entrances to the burrows in which it spends a large part of the year.

Fig. 19.1 Beaver dam (foreground) with adjacent beaver pond. The dam pond is still intact, but from the background of the picture (mouth of the stream) and from the edges it is already heavily silted up and formed as a beaver meadow with grasses and tall shrubs. The photo comes from the northeastern USA. (© Markus Auer)

Fig. 19.2 A picture from the Eifel showing a later stage from a comparable perspective. The beaver dam (foreground) has largely dissolved, the former beaver pond behind it has completely silted up. Woody plants are encroaching on the tall shrubs, marking the beginning of reforestation. The beavers have long since migrated and dammed the stream elsewhere. (© Christof Angst)

The beaver dam and the beaver pond already create **new habitats** which would not exist in this form without the beaver and which are colonised by other plant and animal species. The gradual silting up of the pond, also due to the washing in of materials by the course of the stream, in co-operation with the wood-destroying nailing activity of the beavers, causes the creation of so-called beaver **meadows,** which can be colonised by light-loving herbs. Without the beaver, both stagnant water and light-flooded habitats close to the ground would not naturally exist in Central Europe outside the areas that have recently been over-formed by ice, either not at all or only in rare geo-ecological exceptional situations. Thus, the beaver creates new habitats that contribute to the overall natural diversity of the overarching forest habitat. This also applies to sites of different nutrient salt enrichment through introduced organic material such as pieces of wood or beaver dung, which provide for an ecological fine differentiation of the habitat beaver meadows.

When the beaver pond has largely or completely silted up, the animals create a new pond elsewhere. The abandoned site then undergoes **natural succession** towards tall forest. The beaver meadows gradually become overgrown, after they may have provided grazing for other game species in the forest for a few years due to their abundance of grasses and herbs. The cycle then begins anew elsewhere.

Agriculture Minister Hauk Wants to get at the Beaver's fur

The previously protected beaver could soon be targeted by hunters. "The beaver population is getting so out of hand that we will have to manage it in the medium term," said Agriculture Minister Peter Hauk (CDU) in Stuttgart. "In doing so, we also have to think about the possibility of setting traps and hunting him that way." According to the ministry, the population in Baden-Württemberg has increased from 1000 to 3500 since 2008. "The beaver is spreading extremely fast because it reproduces insanely fast."

[…].

The Ministry of the Environment sees the beaver issue quite differently. "The activities of the beaver increase the structural diversity of water bodies and thus also the diversity of species and the self-purification capacity of flowing waters," said spokesman Frank Lorho. The development of the beaver population is "a success for the protection of species—not only for the beavers themselves, but also for other species that can settle with higher structural diversity". He also pointed out that there were 20,000 beavers in Bavaria. (Stuttgarter Zeitung v. 3.1.2017).

The reintroduction of the beaver was certainly facilitated by the fact that nobody is afraid of these animals. This is less true for the large herbivores, the elk *(Alces alces)* and the **bison** *(Bos bonasus).* Both species are no longer to be found in the wild in Germany today (with the exception of a few settled and immigrant specimens), especially as the preferred extensive lowland forests are largely absent. A regrettable circumstance here is that most of the large-scale nature reserves in Central Europe, especially national parks, have been created either in the coastal area or in the low and high mountain ranges.

However, attempts to settle the large predators **brown bear** *(Ursus arctos)*, **wolf** *(Canis lupus)* and **lynx** *(Lynx lynx)* meet with particular resistance. All three are feared as competitors by hunters, but the former two are also feared by the population at large, which wants to recreate in the forest. Attacks by wild brown bears on humans are rare, but have been documented in isolated cases. Whether wild wolves ever attack humans at all is at least considered controversial.

In this context it seems downright embarrassing when in the 1980s the then German Chancellor Helmut Kohl went on a state visit to Brazil and there vehemently promoted the protection of the jaguar, among other things, but at the same time a conservation initiative for the reintroduction of the lynx in the Black Forest failed due to the joint resistance of farmers and hunters.

If we now have small populations of wolves in eastern Saxony and Brandenburg, for example, and if isolated lynx scour our forests, this is not due to their consistent protection in Germany itself, although both species are protected on paper. In this case, Germany benefits from the much stronger protection in neighbouring European countries for highly mobile animal species that can easily cover several dozen kilometres in one night.

The Wolf: Dangerous or Harmless for Humans?

Red Card for Isegrim

While the wolf *(Canis lupus)* enjoys EU-wide protection, resistance is forming in parts of Brandenburg and Mecklenburg. There, arguments against the animal have been collected for 4 years. A discussion with wolf supporters takes place thereby only conditionally. Members of the initiative "No Wolf" recently organized a discussion round on the topic of wolves in Alt Daber. "We want to draw attention to the fact that the wolf is a danger and the settlement is a mistake. Thus, nature conservation loses its good reputation", initiator Gerd Steinberg from Boek in Mecklenburg underlined the goals of "No Wolf". The 72-year-old used to work full-time in nature conservation. At the beginning of his 45 min lecture Steinberg condemned the media: "They are only available to the wolf supporters and not to us. That's why advocates aren't allowed to make much of a statement today." He reported wolf attacks in the border region between Mecklenburg and Brandenburg and clarified: "We have no intention to eradicate the wolf, but it is not appropriate in Germany" (*Märkische Allgemeine*, 30.1.2012, quoted from Bundesamt für Naturschutz 2012b, p. 193).

Bund Naturschutz on the Subject of Wolves

Wild wolves are very shy animals and avoid contact with humans – even when searching at night for easily accessible food near human settlements. This shyness towards humans is also due to centuries of intensive hunting. Walkers, cyclists and joggers will usually not get to see wolves. The wolves notice humans early and look for the distance. They do not see him as prey[…].

Wolves, like our dogs, are primarily carnivores. They are not specialized on any particular animal species, but hunt what lives in their territory. Therefore, they also attack farm animals and pets. But solutions can be found for this[…] One thing wolves certainly do not eat: humans, no matter what age. Humans do not belong to the prey pattern of the wolf (https://www.bund-naturschutz.de/fakten/artenbiotopschutz/wolf/mythen-und-maerchen-wie-gefaehrlich-ist-ein-wolf.html).

If, however, a brown bear as an animal individual does come into the focus of the media, everything that has already been said about animal protection (Chap. 3) applies to it. And yet the wild brown bear "Bruno", finally released for shooting, could never achieve the sympathy values as the captive polar bear "Knut". A calumny pokes fun at this, but also illustrates it drastically:

Question: Why did Bruno have to die and Knut didn't?

Answer: Because Knut was "white".

Especially in the case of the large carnivores brown bear, wolf and lynx, it is emphatically shown that **hunting** was the main reason for their extinction. Since all three species were always seen as enemies of the farming people (in this case as farmers, shepherds and hunters), they have always been exposed to a special hunting pressure. Throughout the Middle Ages and the early modern period, however, this did not harm the population of these animals. It was not until the nineteenth century, when on the one hand weapon technology improved, but above all when a comprehensive forestry service in the personnel unit of woodman and hunter took care of the stock of the forests and their animal occupants, that the last animals of these three species were shot in the German forests. They remained extinct for a long time until single specimens migrated again from neighbouring countries in the past decades.

A particular success story for nature conservation is the population development of many large birds. Traditionally, these are divided into cultural followers and cultural fugitives. However, some illustrative examples show that probably not a single animal species fears humans per se as long as it is not persecuted or deprived of its living opportunities. Especially in the case of large animals, hunting has often contributed to their extinction. At the end of the nineteenth century, for example, the **Common Raven** *(Corvus corax)* disappeared from Germany, while at the same time the same species had its breeding grounds at the Tower of London, right in the heart of what was then the world's largest metropolis. Until the 1970s, endangered large birds such as black grouse and birds of prey could still be hunted, even when they were on the verge of extinction. Even today, the lobby of hunters is very influential right into the nature reserves and they are generally even allowed to pursue their hunting craft there, while picking flowers that would otherwise be considered weeds is strictly forbidden precisely in nature reserves.

However, large birds are not always hunted and thus pushed back. The **white stork** *(Ciconia ciconia)* is an example of a species for which almost no "natural" occurrence is known, which breeds almost exclusively on human structures (Fig. 19.3). Per se, the white stork is not "nature", but part of rural "culture". It has always been tolerated, if not encouraged, by humans. The popular belief that storks

Fig. 19.3 Tradition meets modernity: A pair of storks *(Ciconia ciconia)* nesting on the solar energy roof of a rural property. Nature conservation and environmental protection are not always so compatible. (Own recording)

make childbearing possible may also contribute to this. In former times, storks were common in our cultural landscape wherever large-scale moors and wet meadows in the lowlands provided for a numerous occurrence of frogs. Due to drainage of meadows and intensification of agriculture, stork populations declined until a low point was reached in the 1980s. Since then, the population has recovered slightly from this low level, helped by the widespread **release of** captive-bred birds into the wild. At the moment, the further intensification of agriculture with the conversion of grassland into arable land (mainly maize fields) poses a renewed threat to the stork population. Similar to the White Stork (*Ciconia ciconia*), the Black Stork *(Ciconia nigra)*, a closely related but altogether rarer species that lives in large forest areas, has seen a slight increase in its population in Central Europe over the past decades.

The reproduction success of the **Grey Heron** *(Ardea cinerea)*, which was pushed to the brink of extinction in Germany by persecution as a "heron" until the 1970s, was much more drastic than that of the storks. Today, the grey heron has become so numerous that it has left its traditional habitat on the banks of bodies of water and can increasingly be found in fields, compost heaps and the like. At the end of the last millennium, the heron's population has grown to a level that should give conservationists in particular cause for concern. Above all amphibians have to suffer from it. The pond newt *(Lissotriton vulgaris)*, originally the most common newt species in Central Europe, has since become rare or disappeared completely. Since the Pond Newt prefers open waters, unlike the two other species also common in Germany, the Great Crested Newt *(Triturus cristatus)* and the Alpine Newt *(Ichthyosaura alpestris)*, it particularly often falls victim to the heron. This again shows that nature conservation cannot protect everything and everyone equally.

What applies to the heron also applied in at least the same form to the **birds of** prey, under which we include both the hawks and the goshawks (Accipitriformes), whose larger forms, again often without closer relationship, are usually called eagles. Between the nineteenth century and the 1970s, virtually all species except the buzzard *(Buteo buteo)* and the kestrel *(Falco tinnunculus)* became quite rare in central Europe. In the meantime, the goshawk, sparrow-hawk *(Accipiter nisus)* and red kite have recovered very well and are widespread again, and the populations of the large species osprey (*Pandion haliaetus),* white-tailed eagle *(Haliaeetus albicilla)* and golden eagle have increased considerably again regionally.

The example of the **golden eagle** *(Aquila chrysaetos)* also shows that there are limits to nature conservation. Since the large animals claim a territory almost the size of a county for themselves, they can never become as common as this may be the case for their prey animals such as the mountain hare *(Lepus timidus),* marmot (*Marmota marmota)* and black grouse (*Lyrurus tetrix*). A further increase leads to the fact that young eagles remaining without a territory involve the existing territory owners almost non-stop in assertion and defence fights, which impairs their rearing success with only one young bird per summer. One could also say that nature itself has ensured, through the largely genetically determined territory size, that the eagles do not become too common and endanger the prey species. This does not seem to apply to the same extent to the food opportunist Grey Heron, which directly competes in its habitat with other species such as White Stork *(Ciconia ciconia),* Black Stork (*Ciconia nigra)*, Purple Heron (*Ardea purpurea),* Buzzard (*Buteo buteo),* Raven Crow (*Corvus corone*) etc.. In any case, at the moment in the Alpine region the suitable Golden Eagle territories seem to be largely fully occupied with a total of several hundred Golden Eagle pairs, less than a dozen of them in the relatively small German part of the Alps.

Among the falcons, the rather rare *peregrine* falcon *(Falco peregrinus) in* particular has increased, and among the **owls** (which, like the falcons, are not closely related to the hawks) this also applies to the **eagle owl** (Fig. 19.4). Both prefer rocky niches, which are very rare in Central Europe except in the Alps and are also frequented by climbers. In addition, the Eagle Owl and the Peregrine Falcon cannot be settled on the same section of rock, as the Eagle Owl likes to claw the Peregrine Falcon. Nature conservation, if one associates certain ideal states of nature with it, must be quite a nerve-wracking thing for the conservationist.

Less fortunate is the development of the **grouse** (Tetraoninae), especially the capercaillie and black grouse. Here it was not only hunting that led to the decline of the large fowl, but above all the **intensification of forestry**. Already in the nineteenth century the populations began to decrease in connection with the tendency to create pure forestry cultures. Dense tree stands and an early timber harvest in the life rhythm of the tree species prevent sparse old-growth stands, which both species need. These old stands provide strong horizontal branches where the capercaillie *(Tetrao urogallus)* retreats to roost and where the capercaillie performs courtship. Light old-growth stands also have an understory of berry bushes (especially bilberry), which are the animals' main food during the warmer months. Since thinned-out old-growth coniferous forests are still more common, especially in the combat zone of the forest at the timberline of the Alps, Capercaillie and Black Grouse have their main distribution in Central Europe there.

Similar to the chicken birds, but not more closely related to them, is the **Great Bustard** (Otis tarda), the largest, at least heaviest bird in Central Europe. Originally a

Fig. 19.4 The eagle owl *(Bubo bubo)*, the largest owl species, is now quite common again in Germany. (Own recording)

representative of the steppe, it has adapted to the **cultivated steppe** of meadows and fields, and is now found exclusively in eastern Germany (and Austria). In recent years there has been a slight increase in the population of this bird, which is acutely threatened with extinction, mainly due to the **reintroduction of offspring.** In the alienation from natural reproduction, nature conservation pays a high idealistic price, which actually contradicts its goals. The modern threat to the Great Bustard is also a consequence of the **intensification of agriculture, in** particular the conversion of grassland to arable land with ever greater proportions of maize, in the climatically already quite dry and therefore in the grassland area non-productive basin and lowlands of eastern Central Europe.

Wild Cattle Roam the Woods Again

They grow up to 1.90 m tall and weigh 900 kg, have a shaggy coat with curved horns and had been extinct in Germany since the sixteenth century: the *bison (Bison bonasus).* Now Europe's largest land mammal is returning. Unrestricted by fences and gates, a herd of eight animals will roam the Rothaar Mountains north of Bad Berleburg in southeastern North Rhine-Westphalia from today onwards.

The new home of the wild cattle is in the forests of Richard Prinz zu Sayn-Wittgenstein-Berleburg, one of the largest private forest owners in Germany. The conditions are ideal for the shy bison: the forest is sparsely populated, virtually uncut by roads and paths, and repeatedly interspersed with meadows and other open areas on which the animals depend as grazers—ideal conditions for the species conservation project that the state of North Rhine-Westphalia and the Federal Agency for Nature Conservation (BfN) have been funding since 2009. "From a scientific point of view, the reintroduction project is important because the bison occupy a currently unoccupied niche of large grass and roughage eaters in the forest that was vacant in Germany after the extinction of the bison and the aurochs," says BfN species conservation expert Uwe Rieken (Benjamin Haerdle, *Stuttgarter Zeitung*, 11.04.2013).

The King of the Night is Back

"Hoo-hoo"—this muffled, somewhat eerie call from the dark of night sends a shiver down many people's spines. But bird connoisseurs are thrilled because they hear the largest owl on earth. For this eagle owl, *Bubo bubo,* with its wingspan of up to 180 cm, was almost extinct in Central Europe in the 1960s. Today, however, the "king of the night" has returned to some regions, in others it is just arriving again. "Bird conservationists currently count 2300 breeding pairs in Germany alone," reports Markus Nipkow, the head of the State Bird Sanctuary in Lower Saxony (Roland Knauer, *Stuttgarter Zeitung*, 17.08.2013).

Cranes Spread Out Again

The Grey Crane continues to spread in Germany and neighbouring countries to the west. About 10,000 breeding pairs were recently counted in Germany—about twice as many as 15 years ago. In its main breeding area, however, the bird is put to tough tests by drought, explained Wolfgang Mewes from the board of the Crane Protection Working Group at the annual meeting of crane keepers in Mecklenburg-Western Pomerania. The birds benefit from the fact that they adapt their way of life to agricultural structures. Many cranes also spend the winter further and further north during milder winters, thus saving energy.

(Stuttgarter Zeitung v. 04.03.2019 n. dpa).

Author's note: The main reason for the increase in cranes is the abandonment of hunting. The additional possible influences of climate fluctuations/climate change on the size of the populations are visible in the report based on the indicated dry summer of 2018 as well as the generally milder winters.

20

Habitats for Land Conservation in Central Europe

Forest Ecosystems

Almost all of Central Europe was originally (and would still be by nature) covered by forest, predominantly mixed **deciduous forest**, in which, depending on the location, the conifers (Coniferales) yew (*Taxus baccata*), pine (*Pinus* spec.), fir (*Abies alba*), spruce (*Picea abies*) and larch (*Larix decidua*) are also mixed in. Only extreme sites such as beach embankments and coastal fringes, young coastal dunes, core areas of upland moors, rock falls, avalanche alleys, boulder heaps or sites above the natural timberline, which in the Alps is generally just below 2000 m altitude (higher in the Central Alps), are naturally free of forest.

However, this uniform forest landscape is not as homogeneous as Tacitus may have seen it in his *Germania* some 2000 years ago. Despite the comparatively manageable number of about 50 native tree species in Central Europe, there are quite different combinations of species depending

K.-D. Hupke, *Nature Conservation*,
https://doi.org/10.1007/978-3-662-66159-8_20

on the location, with two to three tree species usually predominating. This difference in location is based on the different conditions of geological subsoil, soil, climate, exposure conditions and slope.

Although one often does not notice this at first glance, especially as a forest botanist layman, Germany's forests, with their astonishing share of more than 30% of the land area, have also been subjected to enormous transformations. Most of these **forests** are in fact **plantations**, i.e. planted or at least planned stands of rather unnatural species composition and stand structure (Fig. 20.1). One of the mysteries of both domestic nature conservation and **forestry**, which is closely linked to it in many respects, is why these forests can be considered forests, whereas similar plantations in the tropics, for example of rubber trees or oil palms, cannot. Such a classification, which can hardly be overestimated in its importance for the self-perception of nature conservation achievements, is largely due to the very successful self-portrayal of the German forestry industry right up to the mass media (Heimatfilme and Forsthaus books of the 1950s and 1960s, Forsthaus soaps of today).

In fact, the forester has always been first and foremost a timber manager, and secondly also a representative of a lower regulatory authority, which even today can issue reprimands

Fig. 20.1 The hall-beech forest, familiar to most Central Europeans as a landscape image, with straight, tall tree columns of strong beeches that are only limbed far above and largely lacking in undergrowth, is a product of forestry with the intention of achieving even-aged stands. It is often compared, romantically exaggerated, to a Gothic cathedral. Although beech is naturally the dominant tree species on many sites, this hall forest has little in common with a natural forest. Nevertheless, several orchid species important for nature conservation thrive on the seemingly bare forest floor, in particular the woodland bird's-eye (*Cephalanthera rubra* and *C. damasonium*), several marshwort species (*Epipactis* spec.) and the nestwort (*Neottia nidus-avis*). (So-called orchid-beech forest; own photo)

and chargeable warnings to forest visitors, assuming appropriate misconduct. This way of thinking, which is influenced by the structures of the authoritarian state, also finds its way into an understanding of nature conservation, which is primarily concerned with preventing and, if necessary, punishing human misconduct towards nature, such as catching a frog or picking protected flowers. The fact that the forester, in the context of his timber management tasks, is often accused of much greater offences against nature, is in any case irrelevant within the framework of this model of thought. A water-filled wagon track may be filled with road gravel the next day: But it is still forbidden to catch the newts in it.

However, this is not to say that the forestry currently practiced is only bad and only harmful for the interests of nature conservation, for example for **biodiversity in the forest**. It is precisely the diversity of different forms of rotation and operational areas, including clear-cutting, that has also made it possible for light-loving animal and plant species to penetrate the forests to a greater extent (summarised by Ellenberg and Leuschner 2010, p. 295). Among these species are many rare species focused on by conservation. In general, however, this increase is due to ruderal plants, which are less valued by nature conservationists (see also Chap. 13). For the most part, these do not require special protection because, as spontaneous generalists, they are not dependent on the forest.

Some other species, on the other hand, are threatened by the conversion of natural forests into plantations. In particular, these are so-called **K-strategists**, i.e. animal and plant species that depend on stable end-states of succession. Many of them require old wood or dead wood, such as tree cavity nesting birds or insects whose larvae live on dead wood. It is precisely these species, which basically need the "primeval forest", that are particularly dependent on near-natural forests, even if some of them are flexible enough to switch to park trees, and in doing so can even disrupt a metropolitan construction project close to the city centre by their mere presence, such as the lynx beetle (*Osmoderma eremita*) *in* the case of Stuttgart 21.

While Central European forests are rather poor in tree species, the ground flora is richer. It unfolds its growth and flowering already in early spring (March, April), before the forest is fully leafy again. On the one hand, this allows almost full incidence of light on the forest floor; on the other hand, the summer months on the forest floor are often too dry, when the densely leafy forest with its large evaporating surface consumes almost all precipitation water and because the trees with their long water conduction systems develop the higher root suction pressure anyway and thus extract water from the smaller plants on the forest floor. For rapid sprouting and flowering in spring, organs storing nutrient salts and assimilates are required, such as rhizomes in wood *anemone* (*Anemone nemorosa*) or bulbs as in the Gagea species (Fig. 20.2) or root tubers as in the larkspur (*Corydalis cava*).

Fig. 20.2 The forest yellowberry (*Gagea lutea*), a "spring geophyte". (Own recording)

Meagre Grassland

In many respects, **rough grasslands** form a contrast to forests. They are open, characterised by small-growing plant forms of grasses (Poaceae) and herbs, they usually have shallow soils with a thin humus layer and little humus accumulation, they are flooded with light and are dry by Central European standards. If one considers the forest to be the natural vegetation in Central Europe, rough grasslands have only developed through the intervention of man and are thus in this sense remote from nature. What makes rough grasslands interesting and indispensable for nature conservation, however, is their richness, especially in **flowering plants**, which regularly outshines that of mature forests and makes rough grasslands probably the **most** species-rich **biocoenosis in Central Europe**, especially in the variant of **calcareous rough grasslands**, which are widespread in southern Germany on the shell limestones (Fig. 20.3) and Limestones of the Upper Jura (Fig. 20.4), partly also on the river gravels of the Alpine foothills. Within a radius of a few dozen meters, one can certainly find 300 significant plant species here; a biotic richness that one would rather expect in warmer climates.

The site conditions of the rough grasslands are, besides the strong incidence of light, mainly characterized by a **lack of nutritive salts. Dryness**, as the largely synonymous term (semi-) dry grassland suggests, usually plays only a subordinate role, especially since it occurs only episodically in the actually humid climate of Central Europe. Most species are not so-called xerophytes (dry plants). In particularly dry summers, such as 2003, it was observed that many characteristic species, such as *Gentiana cruciata, were* drought-damaged or did not flower at all due to lack of water. However, this is made up for in the normal wetter years by stronger growth and successful reproduction. The term dry grassland refers rather to the brownish-yellowish surface

Fig. 20.3 Walker in the calcareous grassland of the shell limestones in the central Neckar region. The shallow site has been maintained in the past by extensive sheep grazing, is now threatened by forest encroachment and is kept open by mowing crews. (Own recording)

impression of the vegetation, in contrast to the bright green fertilized economic meadows and pastures in the vegetation period. Dry grasslands in the narrower sense are sites without extensive vegetation with patchy vegetation cover and bare areas in between, often overlying rock or slope debris. Here, most plants lack fine soil content in the soil and dryness indeed often becomes the main problem. The **closed rough grasslands**, on the other hand, which predominate by far in the area and are particularly interesting for nature conservation due to their high species richness, exhibit the characteristics discussed above and form an area-wide vegetation of mainly herbs, less grasses.

Calcareous grassland differs floristically fundamentally from **silicate grassland**, which can form on slope debris over granite or on dune sands (here as sandy grassland; Fig. 20.5) and which is usually somewhat less species-rich than calcareous grassland. Many of the plants found here are also old immigrant species (archaeophytes).

Fig. 20.4 A special subtype of predominant rough grassland is the juniper heath, which is characteristic of the traditional landscape of the Swabian Alb, for example. The forest-destroying activity of sheep grazing has left the prickly juniper bushes, which are inedible for sheep, standing here. Following the sharp decline of migratory sheep herds, these are now mostly maintained by maintenance crews that prevent natural reforestation. (Own photograph)

The climatic equivalent of rough grasslands is the **steppe regions** of south-eastern Europe; some of the species, such as most of the orchids represented here, have also migrated from the Mediterranean region in the post-cold period. All of them have the characteristic to disappear immediately in case of stronger shading. However, this does not require a tall forest, shrubs or even just tall herbaceous plants are sufficient. While shrubs like blackthorn and dogwood mark the preliminary stage of forest development, tall herbs are rather restricted to eutrophic (= nutrient-salt-rich) sites, as they need a good supply of nutrient salts for rapid growth until late summer. Examples of such perennial or **tall** herbaceous vegetation are stands of stinging nettle, Indian knapweed (*Impatiens glandulifera*), Japanese knotweed (*Fallopia japonica*), Canada goldenrod (*Solidago*

Fig. 20.5 Sandy grassland near Heidelberg. As the pines left standing for landscape aesthetics reasons make clear, this is also a natural forest site. (Own recording)

canadensis) or giant hogweed. Many of these are neophytes. But it is not so much these themselves as the shift in site characteristics towards eutrophication (for example, through external nitrogen inputs from automobile combustion engines) that is the actual cause of the displacement of native species.

However, rough pastures also represent relics of an originally widespread **extensive farming method**, in particular of sheep pastures, but also of single-cut rough pastures. These were often located on the edge of the village boundaries, far away from the village, and were originally often common land in common ownership. Already in the nineteenth century these areas were in sharp decline, due to the transfer of the common land to private or communal ownership. Often, these agriculturally unproductive areas were either converted into grassland or fields through fertilization or afforested. In any case, however, their transformation was associated with a loss of biodiversity.

In the twentieth century, this decline of the rough pastures was intensified by the extensive abandonment of transhumance. As a result, today mowing crews usually have to be used to keep the rough grasslands open to tree and shrub growth. This change in maintenance methods has also led to a shift in the species spectrum. Plants with thorns, such as silver thistle and thorny knapweed (*Ononis spinosa*), and those with essential oils, such as gentian species, which are rarely eaten by sheep, became rarer, while tread-sensitive plants such as orchids often increased (Chap. 7). Above all, however, this created a maintenance cost problem for official and private nature conservation. In addition, every maintenance operation points to the intrinsically unnatural character of these areas and therefore creates a problem of legitimacy for nature conservation.

Hot Spot Meadow

A closer look is needed, into the small, the tiny—and the view of the big wide world changes. This is the conviction of botanists around Bastow Wilson from the University of Otago in New Zealand. In the "Journal of Vegetation Science", they reported on a study in which they wanted to find the most species-rich region in the world. Commonly, that's tropical rainforests—with their thousands of niches, tiny ecosystems that can consist of just a small puddle formed in the leaves of an epiphyte that serves as a delivery room for a frog. "We looked at the literature on the occurrence of plant species in certain areas," Wilson explained. "Of course, tropical rainforests in Costa Rica, Colombia and Ecuador have the most plant species for areas larger than 50 m^2." But at smaller scales, he said, the picture changes: that's when meadows suddenly become the "hot spots of the planet." According to this, especially in eastern Europe, from the border of Germany to Romania, a particularly large number of different plant species feel at home in a small area (Die Welt, quoted from Bundesamt für Naturschutz 2012c, p. 277 f.).

Author's note: Here one should probably speak of rough grassland rather than meadow (Figs. 20.3 and 20.4).

Meadows and Pastures

Of course, rough pastures are traditionally also grazed. However, this is done very extensively and without maintenance, yield-increasing fertilization measures. However, it is precisely this additional **fertilizer addition**, relatively indifferent whether by slurry or by technical artificial fertilizers, that is detrimental to species diversity. A few particularly vigorous grasses (Poaceae) and herbs, such as the dandelion (*Taraxacum* spec.), increase enormously in number of individuals and growth rate, shade out and displace the majority of less vigorous species. These are then so-called "**fat meadows**" (also called "economic meadows").

The more fertilizers are applied, the more the number of species decreases, especially of flowering plants, but also of insects. Weakly fertilized one- to two-mown meadows (Figs. 20.6 and 20.7) or extensively grazed areas

Fig. 20.6 Species-rich, sparsely fertilized summer meadows in the Glemswald west of Stuttgart. (Own recording)

Fig. 20.7 Special case of orchards. These have developed mainly since the nineteenth century, when many farmers found employment in industry and continued the farm as a sideline. For reasons of working time, they were interested in an extensification of the use in comparison with the until about 70 years ago still very labour-intensive arable farming. Orchard meadows still surround many settlements, especially in the southwest of Germany, as a more or less wide belt. They are particularly endangered by new housing developments. Orchard meadows are no longer profitable today, as the production of both fruit and green cuttings in mixed stands can hardly be further mechanised. Today almost nobody harvests the fruits (mainly apples, but also pears, plums and sweet cherries); apple production has long since switched to the labour-efficient low-stem varieties. The only use still occurring today generally consists of the use of the fallen fruit for the production of sweet cider and fruit brandies. Orchard meadows are important breeding and habitats for a number of bird species such as several woodpeckers (green woodpecker, grey woodpecker, wryneck) as well as for the little owl. In addition, they are valuable for their aesthetics, just as the small-scale alternation of woodland and open land is generally highly valued (Chap. 2). (Own recording)

(Fig. 20.8) can certainly represent target biotopes for nature conservation. However, the trend in agriculture towards intensification and increased yields is moving in the opposite direction.

Fig. 20.8 Grassland in the nature reserve of a former Hudewald is kept open by grazing fallow deer (*Dama dama*). (Own recording)

The correlations between fertiliser application and species richness described above apply equally to **meadows** and **pastures**. Both **grassland biotopes** are also declining due to the increase in maize cultivation, which is usually more profitable or even easier to mechanise. Pasture systems are being further reduced by the increasingly efficient alternative of keeping cattle indoors all year round, which is easier to automate than pasture management and thus reduces labour costs. Feeding is then often via grain or maize silage. **Traditional pasture management**, as it is still practised today on alpine pastures (cf. Fig. 20.14), is particularly favourable from the point of view of landscape aesthetics due to the alternation of open grassland with shade trees, but also benefits biodiversity and thus a second objective of nature conservation.

According to the Federal Nature Conservation Act, the conversion of grassland is prohibited both on (former) moorland sites and in the floodplain of river floodplains. In addition, farmers have already been paid grassland premiums since 2011 as part of the Common European Agricultural Policy (CAP), which have been integrated into the comprehensive greening programme of linking EU agricultural subsidies to ecological requirements since summer 2014. However, the corresponding EU requirements and incentives have so far at best slowed down the general conversion of grassland to arable land, but have not been able to stop it.

A number of nature reserves are now experimenting with grazing by large mammals such as deer (*Cervus*, *Dama* spec.), bison, wild cattle and wild horse backbreds; see also Fig. 20.8). Especially in the vicinity of larger cities, **extensive grazing** that maintains species-rich grassland can be combined with the attraction of "wild animals".

Grassland biotopes such as meadows, pastures and rough grasslands only developed in Central Europe after the forest was pushed back by the first arable farmers and livestock keepers, beginning more than 7000 years ago. Most species migrated from the natural steppe areas of south-eastern Europe, other species originated from the Mediterranean region or from the high altitudes of the Alps above the timberline (Figs. 20.13, 20.16 and Table 20.3).

Moore

According to the geological definition, **moors** are areas that stand on a deep accumulation of raw humus (peat). In the geobotanical and landscape-ecological sense, one will assume permanently wetted areas, which are occupied by a corresponding vegetation and with which, in turn, a characteristic fauna is associated. For nature conservation, this botanical-ecological definition is the more viable one.

However, this does not mean that all bogs are the same or similar. In addition to waterlogging, the supply of nutrient salts is also important. For this reason, a distinction must be made between fens (= lowland bogs) and raised bogs, with the special form of intermediate bogs in the transitional area between the two.

Fens are influenced by groundwater, which has reached a level mostly close to the earth's surface. Often fens are located in the floodplain, the groundwater level is therefore in contact with the river water level and rises and falls with it. As a rule, fens originated from riverine riparian forests and became established when the forest vegetation was removed by humans.

However, the **groundwater** not only supplies water, but also nutrient salts that originate from the mineral subsoil; especially in the floodplains, there is often an additional enrichment with wastewater from the river water. Therefore, the nutrient salt supply of the fen plants is mostly good, although it varies depending on the situation. Depending on the parent rock, a distinction is made between calcareous fens (e.g. on Jurassic rocks of the Swabian Alb or on moraine debris in the Bavarian Alpine foothills) and silicate fens (e.g. in the Black Forest on red sandstone, gneiss or granite). As is the case for more aridity-dominated rough grasslands, the flora is also quite different here depending on the lime content and the lime-rich biotopes tend to be richer in species.

Otherwise, however, the nutrient-poor sites (especially nitrates and phosphates are of concern) are generally much richer in species. Where, mostly due to human influence such as agricultural fertilizers as well as sewage, tall perennials such as cabbage thistle and meadowsweet come up from early summer on, most plants are deprived of life opportunities by shading.

The agricultural pressure of earlier centuries led to a reduction of riparian forests along the floodplains, which were characterised by groundwater and flooding, and to the formation of extensive meadows of moor **grass** (Molineta). Away from the rivers, so-called fen forests in comparably flat locations or in depressions have been replaced in a similar way. In addition to waterlogging, relative lack of nutrient salts and extensive land use have proven to be equally site-determining. Often such meadows are only one-mowed. There is a characteristic succession of vegetation aspects during the year. In early spring, the sequence begins with the flowering of the weakly growing primrose, reaches its peak in early and mid summer (see also Fig. 20.9) and ends in late summer with the taller perennials, which must first develop their large mass through photosynthesis during the warm period. However, the plant mass is mostly dominated by grasses, with the more closely related sedges (Cyperaceae) being important in addition to the grasses. Sedges in particular are difficult to feed to livestock and were mostly used as bedding in cattle sheds. Due to a change in farming practices, moor grass meadows have often become worthless and have been abandoned. This in turn challenged nature conservation with management measures; for the species-rich moor grass meadows threatened to disappear under a willow or alder pioneer forest that was more monotonous in terms of species spectrum.

Raised bogs (Figs. 20.10 and 20.11) are fundamentally different from fens in terms of their site and ecological situation. However, they regularly develop from fens when a climate prevails that is rich in rainfall throughout the year and rather cool in summer due to humidity. In addition, as in the case of fens, flat surfaces are usually required for their extensive formation. These conditions are found in large parts of northwestern Germany and in the foothills of the Alps, as well as on the plateaus of some low mountain ranges.

Fig. 20.9 Marsh gladiolus (*Gladiolus palustris*), a target species for nature conservation in Upper Bavarian moor grass meadows. (Own recording)

Fig. 20.10 Near-natural raised bog with alternation of depressed (wetted with Sphagnum mosses) and raised places with *Carex* spec (sedges). (Own recording)

Fig. 20.11 Drained upland moor in the Bavarian Alpine foothills tending towards heath due to encroachment of heather (*Calluna vulgaris*) and other dwarf shrubs, but also mountain pine (*Pinus mugo*). (Own recording)

Raised bogs are generally no longer found on mineral soils, which is often the case with fens. They are also not connected to groundwater, which is enriched with minerals. Raised bogs are only directly dependent on rainwater. This is due to the fact that they stand on **layers of peat**, which constantly grow upwards due to dead plant matter and can rise several metres above the general land surface. Contact with the mineral aquifer at greater depths was lost with the formation of peat, most of which began at the end of the last cold period about 10,000 years ago. The rainwater is held by the spongy peat fibres and can survive dry periods. However, it provides virtually no nutrient salts, since its formation processes make it largely equivalent to distilled water. Nutrient salt inputs occur naturally to a small extent, for example, through pollen input and insect flight, and in more recent times, anthropogenically increased, for example, through nitrogen oxides. Basically, however, raised bogs are habitats in which, apart from **waterlogging**, the absolute **scarcity of nutrient salts** plays the main role, more than in any other type of vegetation presented so far.

Accordingly, the vegetation is characterised by extremely undemanding plants such as the *sphagnum* mosses (*Sphagnum* spec.). Only here are there also several species that have taken a very idiosyncratic path of additional nutrient salt supply: the "carnivorous" **insectivorous plants**. These include, for example, the sundew (*Drosera* spec.), which is represented by several species in Central European raised bogs.

Raised bogs largely develop naturally in areas suitable for them. Humans established and spread in Central Europe after the ice age at about the same time as they did. At the same time, they probably avoided the core areas of the moors at the beginning. The low nutrient salinity of the habitat made it poor in game and also unsuitable for the later agricultural use. On the contrary, the repression of the forest by the beginning of agriculture a little more than 7000 years ago as well as the destruction of the forest by charcoal extraction has rather led to a further expansion of the raised bogs at the edges, since the thus missing evaporation by the trees leaves more precipitation water for the soil.

It was not until the nineteenth century, and later than in the case of the fens, that large-scale **cultivation of the raised bogs** began, which requires liming to enrich the nutrient salts. This increases the naturally very low pH value (acidic environment) and thus enables the organic decomposition of the peat subsoil with a release of further minerals. Thus, raised bogs can become productive and intensively used agricultural land through cultural measures.

Attempts to convert raised bogs into forest areas by **drainage**, often combined with soil liming, were also widespread. Almost every raised bog in Central Europe has such drainage ditches, which are often dammed or deconstructed in nature-protected areas, resulting in a rewetting effect.

Besides such cultural measures, **peat extraction** has been a major problem for the raised bogs. The history of peat extraction stretches from fuel extraction for the salt works in Lüneburg, for example, from the Middle Ages onwards, through its use for domestic heating systems, to potting soil extraction, right up to the last few decades. Industrial peat extraction, which is hardly ever carried out in Germany, can completely destroy a raised bog. Peat cutting by farmers, on the other hand, has created peat pits in which pools and small ponds have created additional sites, often with a connection to the mineral subsoil that is untypical of raised bogs and thus with species that are alien to raised bogs.

However, in general terms, drainage ditches and peat cutting have led to drying out of the raised bog by run-off waterlogging, and have facilitated an overgrowth of dwarf shrubs such as heather (*Calluna vulgaris*) and the vaccinia cranberry (*Vaccinium vitis-idaea*), bilberry (*Vaccinium myrtillus*) and bilberry (*Vaccinium uliginosum*) playing the main role (Fig. 20.11).

Both due to natural conditions and as a result of cultivation measures, the raised bog is a very diverse biotic community. An ecological gradient runs from the edge of the raised bog to its centre, which is characterised by increasing waterlogging, decreasing pH value and decreasing nutrient salinity.

According to the Federal Nature Conservation Act, peatlands are generally protected, whereby most of the areas that are still considered relatively intact have been transferred to nature conservation areas anyway. In the past, the large remainder mostly fell victim to agricultural intensification, whereby this occurred in several historical waves from the mercantilism of the eighteenth century to the time of the general food shortage following the Second World War.

Bogs have a large spatial distribution by nature, both as raised bogs and as fens. This has been greatly restricted by the influence of man. The moors that exist today have also been considerably altered by cultivation measures. Nevertheless, below the high mountain border, peatlands, along with some forests, have remained among the more widespread habitats those with the relatively greatest closeness to nature (Figs. 20.9, 20.10 and 20.11).

Still Waters and Water Banks

To the inner eye of the observer, water bodies are always special features within the continuity of the landscape, and they are ecologically so too. What makes watercourses particularly species-rich, however, is their pronounced zonation across an increasing gradient of wetness from the "dry" higher watercourse edge to the deep water. However, water bodies also differ in their nutrient content, whereby, in addition to nitrogen compounds, the phosphate content in particular seems to play a greater role here than in "terrestrial ecosystems" (Schönborn and Risse-Buhl 2013).

Depending on the water depth and, above all, the constancy of the water flow, different types of stillwaters can be distinguished (see Table 20.1), each of which allows or prevents different species or biotic communities. Particularly in the case of ponds and lakes, which tend to have little fluctuation in water level, a pronounced zonation of the vegetation is usually recognisable; this is most extensive in flat landscapes (Table 20.2 and Fig. 20.12). It ranges from reeds to a belt of floating leaves to the zone of submerged plants. These belts gradually shift towards the middle of the lake over the course of decades or centuries as a result of a natural process of sedimentation with increasing sedimentation of the substrate and thus decreasing water depth, which can lead to the rapid disappearance of the entire body of water, especially in smaller and nutrient-rich (eutrophic) lakes. In addition to this natural process, stillwaters are also endangered as filling areas for excavated earth and by nutrient enrichment (eutrophication), in the case of smaller bodies of water mainly from the surrounding intensive agriculture. In addition, larger bodies of water in particular are also sought-after by the population as recreational and, in particular, bathing sites, which can lead to conflicts of use.

In the natural landscape, stillwaters are rare in Central Europe outside the young-glacial landscapes (north-eastern Germany, Alpine region with Alpine foothills). However, they have been created again and again by humans for various reasons. In the Middle Ages, for example, it was widely

Table 20.1 Types of standing freshwaters

Type of water body/features	Puddle	Pool	Ponder (in case of artificial plant also: pond)	Lake
Water constancy	Irregular/ephemeral (a few days to a few weeks); after heavy precipitation or when snow melts	Regularly; some months, occasionally for years; nevertheless periodically to episodically drying up again and again	Perennial (continuous)	Perennial; usually traversed by flowing water, stabilizing the water table, or with connection to groundwater levels.
Max. water depth	Indeterminate (mostly flat)	Indeterminate (mostly less than 1 m)	A few decimetres to below 4 m (lower limit of light-requiring underwater plants)	Over 4 m
Living world	Protozoa; with longer water duration also mosquito larvae	Aquatic fauna and flora, which must however endure occasional drying out	Aquatic fauna and flora	Aquatic fauna and flora
Importance for nature conservation	Low	Amphibian conservation, but also other groups such as dragonflies, water beetles and water bugs	In the case of "old" ponds, a large number of animal and plant species are the focus of nature conservation, including water lilies; otters, bats, water birds, reed warblers in the siltation zones, pond turtles, green frogs.	(In the case of reservoirs rather low, since a near-natural zoning of the banks is prevented by the strongly fluctuating water level); otherwise like "pond"

Source: Own listing

Table 20.2 Zoning of the lakeshore vegetation of a lake

Designation of the zone, belt/properties	Reed zone	Floating leaf belt	Zone of underwater plants
Water depth	Waterlogged subsoil or shallow water up to approx. 1 m	Approx. 1–2 m	Approx. 2–4 m
Phenotype	Dense stands of plants, grasses or grass-like appearance, as small sedge grassland often only a little dm, but also up to more than 2 m height	Floating leaves lying flat on the water; also true floating plants	Often dense metre-long underwater plant stands with elongated shoots, leaves often finely filigree
Typical representatives	Reed, cattail, sedge, in between also frogspoon and arrowhead (Alismataceae)	Sea and pond roses, Pondweed, floating pondweed	In particular also many pondweed species (*Potamogeton* spec.), Stoneworts (Characeae)

Source: Own representation

Fig. 20.12 Sedimentation zones at a pond in the foothills of the Alps in Upper Bavaria (eutrophic): alder scrub forest in the background, reed zone in the middle of the picture, floating leaf belt with the yellow pond rose (*Nuphar lutea*) *in the* foreground. (Own recording)

forbidden to eat meat during Lent. Since fish was not considered meat, almost every village had one or more fish ponds. Even monasteries, still visible today in Maulbronn in Württemberg, were often surrounded by several ponds or lakes. Today, in the now Protestant regions, most of these were abandoned in the eighteenth and nineteenth centuries and converted into more productive farmland. In the still Catholic Alpine foothills (Upper Swabia, Upper Bavaria, Austria), however, there are still quite a few of these former fish ponds. Due to their often old age, they usually have a high biodiversity, as new species have to be introduced very gradually, often via the approach of water birds. Here, once again, we see the familiar pattern that nature conservation today, contrary to the semantic core of the term, primarily seeks to preserve traditional structures of use, i.e. cultural landscape conservation.

The focus of the previous presentation was on the conditions of rather eutrophic (nutrient-saline-rich) or at least mesotrophic waters. In addition, nutrient-poor (dystrophic) stillwaters on mineral substrates are of particular value for nature conservation. These are only conceivable in forest areas without nutrient salt input from agriculture. They possess a number of species that are otherwise completely absent. For example, in Feldsee, a naturally formed cirque lake in the Black Forest below the Feldberg peak, two species of Brachsenkraut (Isoetes) occur. Both species exist in Baden-Württemberg otherwise only in the neighbouring Titisee (Sebald et al. 1993).

Flowing waters have been omitted from this chapter; a separate thematic focus is devoted to them in Chap. 23. However, since they flow at different rates, their ecological characteristics are also "fluid". Lowland streams, especially those of the North German or Pannonian Plain, are thus often more similar to stillwaters in terms of living conditions and species spectrum.

High Mountain Ecosystems

If you are still looking for "real" natural areas in today's Central Europe, you will only find them in the **high mountains**. Both the extreme climatic conditions with annual mean temperatures below 0 °C as well as steep slopes and the extensive lack of a topsoil largely rule out agriculture and forestry. From an economic point of view, the high mountains are today of interest almost exclusively from the point of view of tourism. In this respect, **tourism and nature conservation in** the high mountains are subject to a certain degree of competition, more so than in other areas. The main disadvantage here is winter sports with their demands for transport links, lifts and levelled pistes. However, due to the steepness of the limestone Alps, only small parts of the high mountains in Germany are accessible for winter sports, in contrast to the Central Alps, which have a flatter geological subsoil.

The high mountains are a very heterogeneous natural area in themselves, which is related to the strong relief development and the associated climatic altitudinal levels. In addition, the slope, exposure to solar radiation and geological subsoil change very rapidly within a very small area, leading to a wide variety of site conditions and thus to a high degree of biodiversity (Fig. 20.13 and Table 20.3). As

Fig. 20.13 Culture and nature in the high mountains. The foreground is close to the settlement boundary in the Albanian high mountains. A vegetable garden is protected against game browsing by a fence. The forest above, as on the opposite slope of the mountain, shows intensive use of wood. Only in the summit area of the mountain range the natural timberline can be assumed, overhung by cirques and steep rocky outcrops. Geologically and climatically, in terms of slope and exposure, but also in terms of the type and intensity of human influence, the sites are extremely varied and therefore also biologically very diverse. (© Wenzel Halla and Stefan Hupke)

Table 20.3 Highly simplified scheme of the altitudinal stages of vegetation in the Northern Limestone Alps

Designation of the altitude level	Natural vegetation	Cultural vegetation/civilizational influences	Approximate altitude NN
Nival level (high-medium-low-nival)	v.a. Lichens and mosses, few flowering plants sporadically in wind-swept places and on steep rocky outcrops. (otherwise snow cover)	*(Provided the relief is not too steep) Winter sports; mountaineering*	Above 2500 m, in the Central Alps Above 3000 m
		Climatic	***Snow line***
Alpine level (high, medium, low alpine)	Alpine Mats (grasses, herbs; especially in the lower area dwarf shrubs)	(*If possible due to the relief*) winter sports; mountain hiking; traditional: mountain pasture farming	approx. 1700–2500 m, in the central Alps approx. 2200–3000 m
		Climatic	***Forest border***
Subalpine level	"Combat zone" of the forest	Winter sports; Hiking; Traditional: Alpine farming	1600–1900 m, In the Central Alps up to 2400 m
Montane level (high, medium, low montane)	Mountain forest as a mixture of deciduous trees (mainly copper beech, sycamore maple) and conifers (mainly spruce, silver fir)	In the upper area winter sports; hiking tourism; traditionally in the upper area alpine pastures, in the middle area intermediate alpine pastures/Maiensässen, below valley farms; Forestry favours spruce forest	Below 1600–1900 m down to the foreland (approx. 600–800 m)

Supplemented by some altitudinal data in the Central Alps; own representation based on Reisigl and Pitschmann (1958), Ellenberg and Leuschner (2010), among others; author's own observations and measurements: Hupke (1981)

the aesthetic effect of the mountains also comes into play, high mountains in Germany are virtually preferred by nature conservation. The German Alps, which are small in themselves, are largely designated as nature reserves.

In the high mountains, different sub-areas can be distinguished. The **mountain forest of the high montane stage** is generally not yet part of the altitudinal stage of the high mountains, which is formally settled with the 1500 m altitude line, according to the vegetation often with reaching the timberline (in the Alps around 2000 m NN). Compared to forests of lower levels, the mountain forest is often somewhat thinned out by nature and culture; the tree crowns are usually small in size in order to offer less resistance to snow pressure in winter. In addition to the deciduous tree species still present, conifers also play an important role in the German Alpine region: apart from spruce (*Picea abies*), larch (*Larix decidua*) and, to a lesser extent, Scots pine (*Pinus sylvestris*) and Swiss stone pine (*Pinus cembra*); in the Austrian Central Alps, larch and Swiss stone pine are the dominant tree species at the timberline. A certain long distance of the trees allows besides high precipitation in the understory a dense vegetation of ferns, dwarf shrubs and herbs. The small-growing shrubs have the advantage of being largely protected against cold and the associated frost-drying by a winter snow cover, and are therefore found at higher altitudes than the trees, as is the case for the alpine roses (*Rhododendron* spec.).

However, not only the flora, but also the fauna of the mountain forests is interesting for nature conservation. This is one of the reasons why the first two national parks in Germany (Bavarian Forest and Berchtesgaden Alps) were designated with a focus largely on the mountain forest zone. The fauna of the mountain forests includes red deer (*Cervus elaphus*), chamois (*Rupicapra rupicapra*), several wild grouse (capercaillie, black grouse, hazel grouse) and golden eagle

(*Aquila chrysaetos*). Some of these species also occur above the timberline, at least in summer, while others are naturally found in the lowlands but have escaped landscape modification and hunting into the often difficult-to-access mountain forests. In addition, remote and extensive mountain forests form potential habitats for European large predators: for the wolf (*Canis lupus*), the brown bear and the lynx, which would hardly be imaginable in the smaller urban and commercial forests of lower altitudes, for example.

The **timberline** comprises a separate altitudinal stage, characterized as the **subalpine stage** according to Reisigl and Pitschmann (1958; Table 20.3). Contrary to popular conception, it is not formed as a clear line, at least under human cultural influences, but as a broad transitional fringe (Fig. 20.14) that may encompass several hundred meters of elevation. The forest gradually breaks up into smaller groups or solitary trees. Often it is the large trees that have been left standing as shade trees for the grazing livestock of the

Fig. 20.14 The high montane and subalpine levels offer a variety of sites: from wooded areas and open pastures to riparian corridors and rocky outcrops. (Own recording)

alpine pastures. The variety of locations from full sun to total shade and the richness of structures make especially the subalpine stage as a "combat zone" of the forest a species-rich habitat. Already the different exposure allows different sites to emerge. Often even more important is the fine differentiation of the terrain. This determines the height and duration of the winter snow cover. Especially snow-swept ridges are often colonized by chamois heath (or also: alpine azalea; *Kalmia procumbens*), a dwarf shrub lying flat to the ground, or bear lichen turf. Here the problem at sub-zero temperatures is the so-called frost-drying, when with strong wind and possibly sunlight the plants lose water by evaporation, but cannot supply it again when the ground is frozen. Better protected are the dwarf shrubs such as the *Vaccinium species* blueberry, bilberry and cranberry or the alpine roses (*Rhododendron* spec.) at the edge of the small terrain depressions (snow valleys), where the snow often remains metres high in winter. In the actual snow valleys, on the other hand, the winter snow cover often remains too long in the warmer season (often until midsummer), which is again unfavorable due to the shortening of the summer growing season. Here, the vegetation is again rather sparse and characterized especially by the alpine bellflower (*Soldanella alpina*), which comes into flower within a few days after the snow has melted and whose flowering dates therefore vary between spring and midsummer depending on the terrain situation and weather.

In addition, there are special sites created by man, such as the larch meadows, where a special abundance of herbs thrives under a loosened and, in spring, still light-flooded winter-bare tree population.

Where the mountain forest has been destroyed (either by man or by frequent avalanches) or where the natural conditions are very unfavourable with bare rock or only little fine soil, mountain pine (*Pinus mugo*) **bushes** spread in the northern limestone Alps (with which we are dealing exclusively in Germany). These dense pine bushes usually consist only of this single tree species and are largely free of undergrowth, at least when densely developed. They are already naturally species-poor communities.

In contrast, the **alpine mats**, which consist of grass and herbaceous vegetation, are very species-rich, especially in flowering plants. In this respect they correspond in many respects to the rough grasslands of the lower altitudes outside the Alps. However, the alpine mats are naturally widespread above the timberline, which explains, among other things, their species richness. However, the species compositions have only developed in the post-cold period over the past 10,000 years, when humans soon began to change the landscape. Here again, one should abandon the idea that nature is always static and that only humans bring about change.

However, human impact has led to a descent of the timberline often by several hundred metres in altitude, thus considerably extending the habitat of the inhabitants of alpine mats. Even far below the natural timberline, alpine meadows or maiensässen have often been created (Fig. 20.15), which in some respects represent an extension of alpine mats, augmented by some species of lower elevation. That low temperatures are less of a requirement for alpine mats than a lack of nutritive salts and unimpeded incidence of light is shown by the very fact that species considered alpine can certainly occur in the foothills, provided these two conditions are met: Thus, stemless gentian and historically even edelweiss (*Leontopodium nivale*) occasionally thrive on the gravel surfaces of the Isar in the Alpine foothills.

A special habitat within the alpine meadows are the storage **meadows**, which are the preferred storage areas of ruminating cattle, especially around the alpine huts. The

Fig. 20.15 Highly situated farms or alpine huts (here in the Bavarian Pre-Alps at Brauneck near Lenggries at about 1600 m above sea level) have today undergone a change of function towards tourism, for which the good transport connection by a road is a prerequisite. Nevertheless, they played and still play a role in the anthropogenic lowering of the timberline through grazing (cattle browsing). They also create open land areas far below the climatic forest line, where light-loving plants of the alpine mats can descend far. The differentiation of the vegetation around the farmstead is essentially due to the varying input of cow dung. Pastures overgrown with tall herbaceous plants have mostly emerged from storage meadows: Resting places of the cattle in often hollow or flat position; in direct farm surroundings also due to the leakage or dumping of slurry. (Own recording)

surplus of nutritive salts allows the development of a characteristic community of tall perennials, such as the White Germer (*Veratrum album*) or the aconite (*Aconitum*) species. This site, which was initially created purely by human activity, is now threatened by the decline in livestock farming and the abandonment of alpine pastures. The natural conditions of a relatively high forest boundary and a largely closed forest stage below (apart from rockfalls and avalanche alleys; Figs. 20.16 and 20.17) are thus tending to re-emerge.

Fig. 20.16 In rock areas, vegetation is only possible in niches. Specialists such as auricula (*Primula auricula*) or the houseleek (Sempervivum spec.) species settle here. (Own recording)

However, this leads to a certain loss of biodiversity below the forest line, because the light-loving species are pushed back. In addition, the forest, which is by no means monotonous from an ecological point of view, is often perceived as monotonous by hikers and nature lovers. For these reasons, too, landscape conservationists try to preserve visual and aesthetic diversity by keeping areas open.

The actual high mountains, however, are the **high alpine** and **nival stages**, which no longer have any extensive vegetation and in which hardly any new species (seen from below) are added. Thus, the number of species is clearly reduced here. At this point it becomes apparent that nature conservation is not only concerned with animals and plants, but often also with abiotics. Firstly, a natural area that is actually still largely untouched is to be protected here. Secondly, it is precisely the extreme high mountains with rock falls (Figs. 20.13 and 20.16), with ridges and jags as well as with snowfields and glaciers that provide splended impressions of nature.

Fig. 20.17 Avalanche alleys and Mur gullies (as in the foreground of the picture) are the result of extreme dynamics of loose masses in the high mountains (here: in the "Albanian Alps"), caused by the steepness of the relief. Larger woody plants are regularly destroyed. In this way, forest-free open land, which can be colonised by light-loving plant species, is already naturally created far below the climatic forest limit. (© Wenzel Halla)

One of the peculiarities of the reception of the high mountains is that, probably as a result of its special "visibility", it has a greater significance in the public perception than it actually deserves on account of its area. Thus, the federal state of Bavaria traditionally defines itself also on the basis of its share of the Alps. In fact, even with a generous delimitation, the Alpine share of Bavaria is hardly more than 5% of the state's surface area. Bavaria as a whole appears in reality as a predominantly flat and hilly country with a larger share of low mountain ranges especially on its eastern borders. The tendency to make the Alps look more significant than they actually are continues in the self-portrayal of the Bavarian capital: Munich is by no means on the edge of the Alps, as even many visitors think; even though you can certainly see the Alps from the tower of the Frauenkirche on a clear day. To reach them, however, you have to travel another 80 km on the motorway and take just under an hour.

Wadden Sea

The German Wadden **Sea** has undeniable advantages for official nature conservation statistics: It escapes more intensive use due to its special situation and it improves these statistics due to its enormous size relation. Even though the Wadden Sea areas by their very nature no longer belong to the mainland and are understandably not included in the area of Germany: Statistically, they always belong to the protected areas within Germany. Similar tendencies to make the protected areas look more generous than they actually are exist inland, where areas adjacent to a nature reserve, such as roads or other areas of no value for nature conservation, are often included in the nature reserve. The official figures of about 4% of strictly protected areas (not including so-called landscape protection areas, which are hardly subject to any protection) are therefore once again clearly put into perspective.

Apart from this, the Wadden Sea is certainly suitable as a natural area to be protected. On the whole, it can be considered to be very close to nature, especially as it is a landscape in constant evolution for morphodynamic reasons, in which the daily change of tide as well as irregular storm surges ensure constant change. The Wadden Sea is therefore not stable in the way that human observers of nature traditionally assume. On the contrary, it has a considerable capacity to reverse human changes in a very short time. Even islands are subject to constant change under the influence of wind, waves and ocean currents.

At a superficial glance, the Wadden Sea often appears monotonous, but a closer analysis reveals a diversity of habitats depending on water depths, current velocities and material grain sizes. In addition, it is exceptionally rich in nutrients, both in the form of living and dead individuals and their remains. This combination of habitat characteristics enables a high species diversity with a high number of individuals.

Worms, crustaceans (see also Crustacea), snails (Gastropoda) and bivalves (Bivalvia) are virtually integral parts of the mudflat substrate. However, marine mammals such as harbour seals (*Phoca vitulina*) and harbour porpoises also visit the Wadden Sea at least frequently or have their actual habitat here. For fish fauna, too, the food supply due to **nutrient salinity** is an ideal alternative to the rather nutrient-saline and individual-poor habitats of the high and deep seas.

The designation of the Wadden Sea as a protected area is in contrast to the still quite intensive use of the area by crab fishing (actually: shrimp fishing). But such conflicts of use also occur in the continental area, for example between nature conservation and the interests of hunting and fishing. Here, as there, they are mostly decided in favour of the use.

A particular concern of nature conservation in the Wadden Sea is the population of **harbour seals** (*Phoca vitulina*) (Fig. 20.18). The amiable animals with the large beady eyes can be used as sympathetic figures for nature conservation and make it possible to build a mental and emotional bridge between animal protection and nature conservation (see Chap. 3). After the harbour seal was almost extinct in the southern North Sea until the 1960s due to hunting and persecution as a fishery pest, populations recovered considerably until the 1980s due to the onset of conservation. Then outbreaks of the distemper virus severely affected the species. In the meantime, however, the population of

Fig. 20.18 Mass gathering of resting harbour seals on an outer sand of the German Wadden Sea. The populations were severely decimated by canine distemper in 1988, but have since recovered well. (© Marion Hupke)

harbour seals in the German Wadden Sea has recovered strongly and has grown to several thousand animals.

The distemper virus in particular revived the discussion about pollutants in the Wadden Sea, which is located in the area of the mouths of large rivers (Weser, Elbe). In connection with environmental toxins, it can come to a weakening of the body's own immune defence. Here, environmental protection and nature conservation actually have a common field of interest. However, as is so often the case with complex environmental phenomena (e.g. forest dieback in the 1980s), the causal links are very difficult to prove.

21

Small Biotopes and Their Importance for Biodiversity and Nature Conservation

A **forest tree** dies. It may have been there for just a few decades as a birch or Norway maple, perhaps a few centuries more as a long-lived tree species such as oak or beech. The death of a tree takes place slowly. Its wood gradually becomes rotten, insect larvae invade it and live off the slowly decomposing material. This again attracts other animals, such as predatory insects like the chrysalis robber or ichneumon wasps, and also bird species, especially woodpeckers. These like to lay their nesting cavities in these old and dying trees. Often the nest cavities are subsequently used by other animals that are not able to create tree cavities themselves, such as tits, bats and dormice. Perhaps the tree will also rot out internally, making it a suitable resting place or breeding ground for wildcats and pine martens. At some point the old tree will lose its stability and fall over, perhaps in a storm. If it is still vital enough at that time, its root plate may rip up the soil, creating a small depression that, if the ground is waterlogged, may fill with water and form a

K.-D. Hupke, *Nature Conservation*,
https://doi.org/10.1007/978-3-662-66159-8_21

short-lived pool. In any case, the fallen tree tears a gap in the stand that allows increased incidence of sunlight on the forest floor and thus new plant life. In addition, it provides great nutrient salt potential with its **decaying wood**, so its death brings lush plant life. Often, germinating spruce trees stand on a decaying trunk as if arranged in a row. In addition, a large number of fungi will grow on it, and more beetles and other insects lay their eggs in its wood and use it for the development of their larvae.

This situation was certainly common until the greater use of the forests by man. Only the **development of the forests**, which in most low mountain ranges did not occur before the High Middle Ages (from about the tenth to the twelfth century), led to a decline of the old trees and thus to their early death. After a rather unsystematic and thus stock-destroying timber use in the Middle Ages and early modern times, modern forestry was developed from the eighteenth century onwards with the aim of sustainability. In this early period, **sustainability** (see von Carlowitz 1713) simply meant that no more wood should be removed from the forest in a given period than could grow back in the same time. This principle was phenomenally new at the time, because it meant that consideration was also given to the utilization demands of future generations. For nature conservation, this early use of the term sustainability is certainly of ambivalent value. On the one hand, the forest continues to exist in its wood mass as a result of sustainable management, or is even being rebuilt. On the other hand, the traditional sustainability definition of forestry does not protect the forest from being structurally simplified and contaminated with species alien to the site, which serve the pure timber yield. The forest tends to be converted into a plantation.

In the sense of an expanded understanding of sustainability, nature conservation in the forest (Scherzinger 1996) is today striving to reclaim areas that are left to themselves in the form of process conservation and are surprisingly rich in species. However, there is a clear shift in the species spectrum in comparison to commercial forests (woodlands): The light-loving species along the now non-existent roadsides and impact clearings largely disappear, and the actual forest species reappear.

In addition to the natural phenomena of **old-growth** and **deadwood**, however, the cultural landscape is dominated by **small biotopes** that were created through human intervention. Most of them did not serve the purpose of increasing biodiversity, and most of them had no pronounced aesthetic value. Those who still work in rural areas today, but do so with an "educationally ennobled" urban gaze, are often downright shocked at the farmers' lack of interest in the **beauty of the cultivated landscape that has evolved**. This beauty has come about as a by-product of use, and it is not even beautiful in itself. Were it otherwise, the farmer would undoubtedly perceive it. Rather, the beauty of both the natural and cultivated landscape is a projection of the urban and educated classes, who have become distant from both nature and agrarian culture in their daily references. Farmers can usually only perceive this beauty of a traditional agricultural landscape when they have distanced themselves from it professionally and spatially, for example, when they have given up farming and moved into an apartment in the city.

Such **small structures**, which stand for the beauty, but also for the diversity of the landscape, can be water catchments, which serve to water the cattle, or field hedges, which separate one's own field from that of the neighbour and thus help to avoid border disputes. Or they can be stacking areas for winter firewood, or a small copse behind the house that provides that very woodcut. Or solitary trees that provide needed shade for grazing livestock in the hot early and mid-summer. All these small structures provide hiding places and food for numerous animal species.

Forest edges combine many of the characteristics described below for field hedges. As characteristic **border and transition areas**, they are particularly rich in species, often more species-rich than the closed interior of the forest or than the often bare field. This species richness, especially in insects and birds, depends on the structural richness of the forest edges. This is especially true for a richly stepped crown structure, a mixture of old and young trees of different species as well as for an oscillation of the forest edge with different exposures and thus again for variations in solar radiation, humidity and temperature regime. In addition, the forest edge benefits from the transition of different habitats. For example, many game species use the retreat space of the forest, but like to seek out fields and meadows for grazing, as is common with roe deer, red deer and also wild boar.

By thinning out the forest and, in particular, by creating a more or less regular alternation of field and forest, man has ensured that forest edges, which by nature would only be found in the forest steppes of, say, south-eastern Europe, have become a very characteristic element of the cultivated landscape.

Individual trees, often used as shade trees for grazing cattle, often develop a very expansive tree crown as a result of the free incidence of light, which usually starts a few metres above the ground. Very strong horizontal branches often dominate the crown, making them interesting as climbing trees for children. The great development in width is often at the expense of height, which, unlike in dense forest stands, is not an absolute requirement when there is also strong lateral incidence of light. In addition to native forest trees, cultivated species such as fruit trees also occur as solitary trees.

Avenues are in some respects a special case of solitary trees with a row, often in a typical double row, along a path or street. Also a special case of the solitary tree is the village lime tree on or next to the village square, where it once marked the geometric centre of a small village.

Larger birds such as birds of prey or crows (Corvus spec.) often build their nests in individual trees. The carrying capacity of the branches of a field hedge would not be sufficient for this. The ecological "harmfulness" of these birds to smaller species as well as to hunting game is generally judged much more cautiously by wildlife biologists today compared to past decades.

In most cases, solitary trees do not belong to the rare species and would not actually be worth protecting for reasons of species protection. However, they are invigorating and aesthetically valuable elements of the cultural landscape, and they are also not to be despised as biotopes and thus habitats for many other species. As Chap. 2 has already shown, nature conservation is nourished by a variety of motivations.

Particularly valuable elements of diversity in agricultural areas are **field hedges** (Fig. 21.1), as they are, and even more were, widespread as curtilages in eastern Schleswig-Holstein. They are richly sunlit and yet shady in the interior. Their structures resemble forest edges and attract species that would naturally be found in the transitional areas of the forest steppes of south-eastern Europe. Small predators such as weasels (Mustela spec.), hedgehogs (*Erinaceus europaeus*) and shrews (Soricidae) shelter here and help agriculture with pest control. Field hedges are also important as breeding sites for dozens of bird species. It is only through field hedges and field copses that the monotonous agricultural steppe can be used at all as a supplementary habitat for most species.

Botanically, field hedges are also quite rich. Characteristic are the hedge-forming shrubs or low trees such as blackthorn, dogwood, barberry, privet, hawthorn or field maple. A longer and multiform hedge is often formed by more than a dozen shrub species. The width of a field hedge is usually only a few meters and corresponds to its purpose of demarcation. Often it also follows a field path. Inside a

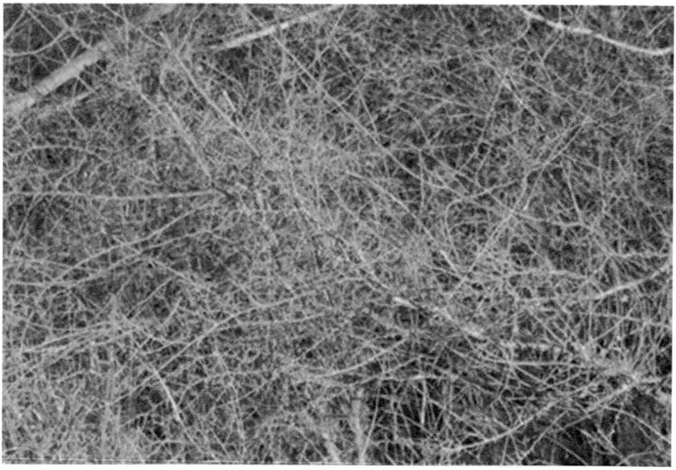

Fig. 21.1 Thorn network of sloes in the small biotope field hedge. The thorny hedge, which is still bare in early spring, allows many small songbirds to breed without danger. (Own recording)

hedge there is often a great lack of light due to the heavy foliage of the shrubs towards the light, so that an actual undergrowth is usually absent. At the edges of the field hedge, on the other hand, the **incidence of light** is so great that the vegetation is similar to that of an arable field or field margin. The strong change in the average incidence of light creates a variety of small and very small habitats, even in the canopy of the hedge, which is usually only a few metres high, and thus a variety of insect species and other arthropods in particular. These in turn support numerous bird species, especially songbirds. The high density of hedges, due to the strong incidence of light from above and from the sides, offers songbirds—partly supplemented by thorny branches, for example of rose species, blackthorn and hawthorn—protection from predators in their breeding zones, while the birds themselves, due to their small size, can mostly move easily between the branches, leaves and thorns (Fig. 21.1).

Dry stone walls (Fig. 21.2) were built on steep slopes, where they help to stabilize slopes and roadsides and create the level areas in steep relief that are necessary for more intensive use. These terraced sites are often combined with vineyards, for which sloping sites are preferred in Central Europe for microclimatic reasons. However, the progressive mechanisation of agriculture has also proved disadvantageous here. Either the former small terraces were transformed into large terraces, as is probably only possible in deep loess sites, especially in the Kaiserstuhl. Or they were abandoned, which leads to increased soil erosion if the use is maintained, as can be observed in the Württemberg Keuperbergland (Stromberg, Heuchelberg, Löwensteiner Berge). The third option would be to leave it to natural succession with the final stage of forest. However, in the Neckarland and in the Mosel region, less pronounced in the

Fig. 21.2 Small biotope dry stone walls in south-west German vineyards. With their crevices and cavities, the dry stone walls provide shelter for numerous amphibians and reptiles such as the Common Toad, Fire Salamander and Smooth Snake (*Coronella austriaca*). (Own recording)

Middle Rhine, viticulture in small terraced sites has been preserved until today, partly quite distinct.

Especially the dry stone walls set in gaps represent ideal retreats for amphibians and reptiles. A few years ago, friends of the author carefully dismantled and rebuilt about 8 m of dry stone wall that had fallen into disrepair on their hobby property in the middle Neckar region north of Stuttgart. In the process, they encountered the following amphibians and reptiles during the winter resting period: Fire Salamander (*Salamandra salamandra*), Pond Newt, Common Toad (*Bufo bufo*), Grass Snake (*Natrix natrix*), Smooth Snake (*Coronella austriaca*), Sand Lizard (*Lacerta agilis*) and Slow Worm (*Anguis fragilis*); in total, three amphibian and four reptile species in only a few meters of length. The fire salamander in particular seems to have left its original habitat in traditional wine-growing areas in the vicinity of damp forest gorges and spring brooks and found its actual habitat in the dry stone walls of the vineyards. Almost nowhere else can it be shown as well as in the example of the dry stone wall that species are quite capable of colonising man-made habitats, and this in impressive numbers and great spatial density. The decisive factor is the stability of structures created in this way over decades, or even centuries, and above all their internal differentiation and diversity. From the outside, dry-stone walls often appear bare and hostile, but one does not necessarily see the biotic richness they literally contain (Fig. 21.3).

Standing waters are naturally subject to a rapid input of sediment, provided they are suitably large and have at least one stream running through them; or they are small and are largely under trees, resulting in sedimentation with decaying silt within a very short time during autumn leaf fall. To put it briefly, apart from the zones of recent glaciation, the natural landscape is generally very poor in standing water. For the most part, these were probably limited to

Fig. 21.3 The Mediterranean Wall Lizard, whose northern limit of distribution passes through southwestern Germany, is a typical inhabitant of dry stone walls. The animal depicted here represents a successful conservation relocation attempt in the context of newly constructed dry stone walls. (Own recording)

strangulated meander bends of flowing waters or to beaver ponds in the Pleistocene non-glaciated part of Central Europe before the stronger interventions by humans.

This is surprising, because many animal and plant species are dependent on standing water bodies or on their riparian margins and siltation zones. In addition to almost all native amphibian species, these include above all many species of aquatic insects (beetles, water bugs, dragonflies, stoneflies and caddisflies). For the life possibilities in such a water body, it is quite decisive whether and how often it dries up. Occasional drying out (**pond** habitat; cf. Fig. 21.4) is often a prerequisite for the animal species. Perennialisation would in fact mean the emergence of fish whose spawn is occasionally introduced by the plumage of waterfowl. Such a fish population would threaten the amphibian offspring. However, if the pond dries out too frequently and cannot

Fig. 21.4 Wagon track ponds as small biotopes. Mechanized forestry often leaves deeply churned tracks, which are criticized by hikers and environmentalists alike. However, the tracks shown provided habitat for the protected amphibians Alpine Newt and Yellow-bellied Toad. (Own recording)

retain its water (**puddle**) for at least a few months, it will be largely devoid of aquatic life, with the possible exception of mosquito larvae. A pond, on the other hand, fills with water mainly in the winter months and holds it more or less regularly until early summer. Such a body of water is ideal for spring spawning amphibians such as newts, brown frogs (*Rana* spec.) and toads (*Bufo* spec.). Even if such a body of water not infrequently retains water only once in several years until the tadpoles can develop, this is often sufficient to maintain the respective population, since the parent animals as individuals are no longer dependent on the pond and amphibians can survive for many years. Many years ago, the magazine *Das Tier* mentioned in an article about fire salamanders that the animals could live up to about 20 years. Thereupon a reader came forward who credibly explained that a fire salamander (*Salamandra salamandra*) caught by himself 52 (!) years ago as a larva was still living with him as an adult animal. This longevity explains that a spring creek that has been completely dry for more than 10 years can suddenly contain fire salamander larvae again when the water flows again.

The frequency of water flow is also crucial for **small streams**. In the hilly areas and low mountain ranges of southern Germany, small headwater streams that often dry out completely as a result of prolonged dry periods are regularly colonized by the larvae of the fire salamander. If, on the other hand, the streambed dries out only irregularly and infrequently, for example about once on average over several years, the stone crayfish (Fig. 21.5), a small species of crayfish, takes the place of the fire salamander larvae. This does not tolerate regular drying, but does tolerate very occasional drying, as it is also mobile overland and can search

Fig. 21.5 The stone crayfish (*Austropotamobius torrentium*) is the smallest of the native crayfish. It inhabits clear upper reaches of streams in southern Germany and Austria and is as sensitive to organic pollution as it is to chemical pollutants. Thus, it can serve as an indicator for unpolluted stream sections. (Own recording)

downstream for new habitats on wet nights. If, on the other hand, the water supply is perennial, the stone crayfish is replaced by the fish species bullhead (*Cottus gobio*), minnow (*Phoxinus phoxinus*) and brown trout (*Salmo trutta*), which are likely to quickly eat its larvae or even the parents. Overall, the frequency of water flow is more important to the fauna than the strength of the mean flow itself. Even a small stream may pour very constantly under certain circumstances, and a larger stream may dry up regularly, depending primarily on the geological substratum of the watershed.

Because **spring outlets** (Fig. 21.6) draw their water from great depths and therefore have the same water temperature throughout the year (namely the mean annual temperature), they form quasi-aseasonal oases in Central European landscapes, which have no seasonality and are therefore

Fig. 21.6 Spring pond as a small biotope. With its special conditions (almost constant low water temperature all year round), this habitat offers opportunities for numerous aquatic insects, but also for newt and fire salamander larvae (both found at the site). (Own recording)

more reminiscent of the mountains of the tropics. Besides the fauna, especially arthropods: caddisflies (Trichoptera) and stonefly larvae (Plecoptera), water beetles, water bugs, this aseasonality also affects to a lesser extent the vegetation of a spring outlet, which is often deepened to form a spring pool. Here, typical representatives such as watercress and bittercress (*Cardamine amara*), water star (*Callitriche* spec.) and creeping speedwell (*Veronica beccabunga*) mostly remain green throughout the year, but only flower in the warmer season outside the spring water.

Since groundwater has undergone a mostly considerable purifying and filtering effect depending on the geological subsoil, water pollution in headwater streams generally plays a minor role compared to larger water bodies or flow sections downstream. Nearly all major rivers begin in their headwaters as pure water in the upper quality categories, as indicated by water quality mapping. A regular problem outside the forest areas, on the other hand, is the straightening of small watercourses, in extreme cases within concrete half-pipes, in settlement areas also often laid completely underground. However, it should not be concealed that, in addition to other sins in this respect, the renaturation of watercourses (see Chap. 23) has now become common practice. This creates diverse structures close to nature and thus a variety of habitats for many animal and plant species (Fig. 21.5).

22

Geological Landscape Objects in Nature Conservation

In most cases, nature conservation is concerned with the preservation of endangered or characteristic animal and plant species; at least this is usually the historically oldest intention. In recognition of the ecological embedding of plant and animal organisms in the surrounding habitat, **biotope protection** became increasingly important. Here too, however, the conservation of endangered species was and still is the main focus. In addition, according to the **Federal Nature Conservation Act**, the beauty and uniqueness of landscapes and landscape elements are to be protected. This intention goes beyond the protection of biotopes and ecosystems and includes abiotic natural elements. This also regularly includes geological objects.

The **geological objects** are manifold. Limestone areas with karst caves, sinkholes, dry valleys, calcareous tuff formations (Fig. 22.1) or, in the alpine region, cirques and hummock meadows are particularly rich in geologically striking small and meso forms. In addition, natural rocks

K.-D. Hupke, *Nature Conservation*,
https://doi.org/10.1007/978-3-662-66159-8_22

Fig. 22.1 Over-rilled limestone tufa rock. Precipitation of lime is due to segregation of carbon dioxide as the spring water escapes from the bedrock. The solution equilibrium of calcium hydrogen carbonate in the water is thus shifted and an excess of calcium carbonate is precipitated around the fine branched moss plants (lime tuff). (Own recording)

and landslide terrain in general are preferred objects of nature conservation. In the volcanic area of the Eifel, the maars represent special target objects. Waterfalls, meander arches and spring outlets can also be counted among the geologically and geomorphologically determined natural objects.

In general, however, it can be said that almost all of these geological monuments also have special site conditions and are conspicuous and rare as biotopes. Thus, it is often not possible to separate which part of the protection is due to geology or geomorphology and which to biology. For example, karst caves form important winter quarters for bats, cliffs are important breeding sites for eagle owls, peregrine falcons, golden eagles and ravens, in the area of the waterfalls are the main occurrences of the otherwise rare naturally protected stag's tongue fern *(Asplenium scolopendrium) as* well as several moss species that require constant soil and air humidity.

In general, it can be observed that nature conservation often favours **extreme areas** (cf. Chap. 5), which are mostly defined in geological-geomorphological terms. A diversity of rare animal and plant species can be found here, if only because the intensity of human use has always been lower on rocks and in bogs, in steep gorges and in the Wadden Sea, which is flooded during the day, than on other areas that are more suitable for agriculture and forestry. However, this lack of use of extreme and rather untypical sites for the large area also facilitates the removal of these areas from use and their protection.

In addition, the **aesthetic appeal** as a criterion for nature conservation cannot be overestimated. This applies in particular to geological features such as rock formations, dunes, sinkholes or caves. For example, the oldest naturally protected object in Germany is the Drachenfels in the Siebengebirge near Bonn. Not infrequently, the factual argumentation seems to be postponed compared to the unscientific aesthetic impression.

More rarely, geological objects have been placed under protection because of their own scientific or educational value. This is especially true for geological outcrops (Fig. 22.2), which per se have almost always been created by humans and therefore tend to be rather unnatural in character. Due to their small spatial extent, geological objects are often designated as natural monuments and not as extensive nature reserves, which, however, does not usually mean any difference in protection status (Figs. 22.1 and 22.2).

Fig. 22.2 Outcrop showing geologic faulting of strata, as a natural monument. (Own recording)

23

River Straightening Versus River Restoration

A **natural stream or river course** (Fig. 23.1) has very different habitats and living conditions in its bed: Shallow and deep water zones, higher and lower flow velocities, different densities of riparian and submerged vegetation, different structure of bottom substrate. The main controlling factor is the **flow velocity**. In addition, depending on the flow situation, there are areas dammed by beaver dams and the oxbow lakes of cut-off river loops. In short, a natural **watercourse** is extremely inhomogeneous and therefore rich in different sites and partial habitats.

But this diversity is under threat. Even in earlier centuries, river riparians repeatedly pierced individual meander bends and thus created course shortenings. This created valuable cultivated land, but also encouraged structural impoverishment in the riverbed. In the nineteenth century, hydraulic engineering progressed. During this period, the correction of the Upper Rhine, initially under the direction of the Baden colonel Tulla, was carried out as the largest

K.-D. Hupke, *Nature Conservation*,
https://doi.org/10.1007/978-3-662-66159-8_23

Fig. 23.1 View of the still unregulated Upper Rhine. (Painting by P. Birmann. View from the Isteiner Klotz up the Rhine towards Basel, c. 1819. Inv. no. 71. © Kunstmuseum Basel; Martin P. Bühler, with kind permission)

overall measure of this kind in Central Europe to date. In addition to **straightening** and structural simplification of the river bed, this also led to the destruction of riverine floodplain forests and the clearing and use of the resulting fens (Fig. 23.2).

What happened on a large scale on the Upper Rhine was repeated on a small scale at countless locations: Streams have been straightened, relocated into pipes to prevent further erosion, and in the vicinity of municipalities they have been completely blocked. In the process, not only near-natural and species-rich biotic communities were lost, but also lively parts of the landscape and play areas for children to build dams and splash around in.

From the 1970s onwards, a countermovement arose to **restore** streams and rivers to their natural state (Fig. 23.3). During a transitional period of one to two decades, it was possible to observe how watercourses located in the same

Fig. 23.2 The Rhine today (with Rhine side channel), only a little to the north and with a comparable perspective compared to Birmann (Fig. 23.1). (Source: Wikipedia.com)

district in spatial proximity were straightened on the one hand and renatured on the other through various public funding programmes. But perhaps such criticism is unfair, after all, there is a paradigm shift in the planning treatment of flowing waters that cannot simply be decreed. After all, it has now become generally accepted that near-natural watercourses are a necessity. Straightened and concreted sections of streams still exist because old structural changes continue to have an effect. But in the course of new redevelopments, such results of spatial planning activities have at least become rare. This means that there is a chance that biodiversity will gradually increase again, even in the area of settlements. The Federal Nature Conservation Act supports this development by providing special protection for vegetation strips near watercourses. These linear structures prepare the way for the planned networking of protected habitats in a natural way, as it were.

The availability of nutrient salts, which has been increased to gigantic proportions by man through household wastewater and fertilization, is a problem especially for naturally oxygen-rich **flowing waters**. This connection has already

Fig. 23.3 Renaturalised stream course. Rock blocks inserted differentiate the flow velocity and, depending on the flow exposure, create their own microhabitats for molluscs, crustaceans (see also Crustacea) and aquatic insects. On the other hand, they also serve as stepping stones over which the watercourse becomes accessible without causing great damage to fauna and vegetation. (Own recording)

been described several times for meadows and pastures, but it has a different significance for bodies of water. Here, the additionally available nutrient salts (**eutrophication**) lead to a mass development of microorganisms, algae and flowering plants. These die off, especially in autumn, and are decomposed by destructors (decomposers such as bacteria, fungi), binding oxygen. However, this oxygen is then lacking for the animal organisms, which mostly breathe the atmospheric oxygen dissolved in the water. This causes mass deaths among them at intervals. In most cases, only species with a low oxygen demand can survive in heavily contaminated waters in the long term. In Germany, there is a measurement system for assessing **the quality of flowing waters** which, in addition to a few chemical measurement parameters such as the content of nitrogen compounds, the content of phosphorus compounds and oxygen depletion, works primarily with the aid of animal indicator species. However, the **bioindicators** used are not so much fish, which are difficult to catch and identify, but the much more species-rich invertebrate groups such as aquatic insects (larvae), crustaceans and molluscs such as snails and mussels. Especially in species-rich animal-systematic groups, the individual species often have only a small characteristic range of tolerance of eutrophication within which they can maintain themselves in their ecological niche in competition with other species.

Today, with the exception of small-scale headwater regions in the middle of forest landscapes, there are hardly any watercourses that can be considered unpolluted or little polluted. In this respect, there has been a shift in the species spectrum towards those species that can use a high level of nutrient salts and that also manage with little oxygen. All other species are threatened, especially in the lower and middle reaches of the river systems.

The extent to which agriculture or (despite sewage treatment plants) household wastewater pollute the rivers the most is still being discussed at the moment. In most small streams, however, there is no sewage treatment plant; at least here, agriculture is almost exclusively the polluter.

In contrast to the general description in Chap. 3, water pollution control is a field of work in which the interests of nature conservation and environmental protection converge on a case-by-case basis.

Salmon Offspring on the Moselle

There are salmon offspring not only in the Black Forest, but now also again on the Moselle: More than 20 years after the first reintroduction of salmon, wild-born specimens have been spotted for the first time in the Moselle tributary Elzbach. "This is good news for nature and water protection in Rhineland-Palatinate," said Rhineland-Palatinate Environment Minister Ulrike Höfken (Greens) in Mainz. The return of the salmon shows the permeability of the rivers. Salmon migrate from their home rivers to the sea and return after one to three years to spawn. Among other things, they are hindered by dams, canals and river beds that have been straightened for shipping.

(dpa-message, n. StZ v. 12.8.2016)

The Largest Floodplain Forest in Central Europe

With a total length of 1091 km, the Elbe is the twelfth longest river in Europe. It rises in the Czech Giant Mountains and flows into the North Sea near Cuxhaven. A variety of habitats can be found along the Elbe: flat beaches, steep embankments and gravel and sand banks as well as dunes and river valley meadows. A very low gradient favours the formation of river meanders and densely vegetated oxbow lakes. In addition, the largest contiguous floodplain forest in Central Europe can be found here.

About 380,000 ha of this varied natural landscape have been recognised by UNESCO as a biosphere reserve "River Landscape Middle Elbe" since 1997. With its floodplains, it is also part of the European Natura 2000 network of protected areas over a length of around 400 km. Many specialised and therefore often endangered species have settled in this landscape, which is also characterised by an alternation of high water and drought. More than 1000 different plant species grow in the biosphere reserve alone – including water nut, Siberian *iris (Iris sibirica)* and many orchid species. Among the 40 mammal species is the endangered Elbe beaver *(Castor fiber albicus)*. The varied landscape provides rich breeding and feeding grounds for the 135 regularly breeding bird species such as the grey heron, black stork (*Ciconia nigra*) or red kite *(Milvus milvus)*.

(Von der Homepage des WWF, accessed 1.5.2013)

24

Nature Conservation in the Forest: Natural Forest—Permanent Forest—Clear-Cutting?

Foresters and **conservationists** (whatever one may understand by the latter) traditionally have an ambivalent relationship.

On the one hand, foresters generally see themselves as conservationists by virtue of their office. To this day, district foresters have lower police rights or rights of order in the forest and can also hand out tickets if a forest visitor violates nature conservation and forest laws. The necessary aura of power is given to them by their status as civil servants, their uniforms and their right to bear arms. The latter is also necessary because the district forester generally also exercises the right to hunt, unless the districts are leased to private individuals.

All in all, the various official tasks of the district forester, such as the exercise of state authority, forest protection, forest use, nature conservation and hunting, are clearly in tension with each other. In this context, only the **use of** the forest, especially through the timber industry, yields

K.-D. Hupke, *Nature Conservation*,
https://doi.org/10.1007/978-3-662-66159-8_24

economic returns; the other functions are more likely to be located in the non-material sphere. Because society needs wood and this is the economic foundation of the forester's profession, the district forester is also first and foremost a wood manager.

This existence as a timber manager shapes the view of most foresters, even more so in the older than in the younger generation. Leaving forests unused, even in small areas, is perfectly in line with the other functions attributed to foresters, such as forest protection and nature conservation, and is also not detrimental to the practice of hunting. Only very gradually is there a tendency, associated with a change of generations in forestry, to open up new tasks for the forester's profession, for example in the context of national park management. But as the attempt to designate a new national park in the northern Black Forest in Baden-Württemberg has just shown, the opponents of national parks regard the forestry scientists at the universities as willing "donors of expertise", whose judgements are usually diametrically opposed to those of the nature conservation experts who are also based at the universities.

Accordingly, models that are more oriented towards the protection of the forest than its usability did not receive their initial impetus from professional forestry or forest science. Initiators of a forest image that is not only and not primarily oriented towards timber use have often been urban nature conservationists outside the profession, but also forest owners. For example, from the end of the nineteenth century, the aristocratic forest landowner Friedrich von Kalitsch established a model forest in what was then central Germany, which the forest scientist Möller (1922) characterized decades later as a permanent forest.

A **permanent forest in** the Möller' sense does not know any clear-cuts. The entire forest area remains forested, and trees are only removed one at a time. The forest is a mixture of trees of different species and different age classes. According to Möller, such a permanent forest is more productive and ecologically stable in the long term than the opposite type of age-class forest.

The **age class forest** is established following a clear-cut, usually by planting trees of a species preferred for timber production, often spruce in the uplands, often pine (*Pinus* spec.) on poor sandy soils of the lowlands, and increasingly Douglas fir (*Pseudotsuga* spec.) overall. Depending on the tree species and the site, the age-class forest becomes ready for harvesting after a few decades and is felled over the entire age-class area. This establishes a new cycle of planting and maturation.

For decades, nature conservationists have clearly preferred the permanent forest to the age-class forest. It corresponds to our model of undisturbed nature (the small gaps left by single-trunk harvesting close up again immediately). But is this really an advantage in terms of nature conservation? Or is it not just (once again) our illusion that we have escaped from human society in the forest and are moving in untouched nature? Clear-cutting (the term is also used metaphorically in other social contexts and has a clearly negative connotation), on the other hand, would very quickly tear us out of our illusion of pure nature.

However, possible criticism of the idea of permanent forest also began to emerge from a completely different perspective: from research into the vegetation dynamics of primeval forest systems, especially in eastern Central Europe, such as the **primary forests** in Kubany (Czech Republic) and near Bialystok (eastern Poland), but also in the Romanian Carpathians or on the western Balkan peninsula (Fig. 24.1). Here it can be seen that even the "primeval forest" tends to sudden collapses of old-growth stands and to a kind of age-class-wise reestablishment. A new cycle takes place via a youth phase, a maturity phase, an optimum

Fig. 24.1 Primeval ancient mountain forests in difficult-to-reach locations in Albania. (Photo: © Wenzel Halla/Stefan Hupke)

phase and an age or decay phase. The primary forest differs from the age-class forest of forestry mainly in that the former lacks old-growth and deadwood stands, since the cycle is interrupted earlier here; otherwise there would be no wood to harvest. Structurally, however, such a primary forest, which in itself consists of a mosaic of different age classes (summarized by Ellenberg and Leuschner 2010, pp. 288–296), is closer to the age-class forest than to the permanent forest. This exposes the naturalness of the permanent forest to some extent as an illusion.

Biodiversity is also one of the essential yardsticks of a nature conservation assessment in forests. **Clear-cutting** tends to increase biodiversity because it creates light-flooded areas that attract typical clear-cut vegetation. Since clearcuts remain open for only a short time, but nevertheless for several years, tall herbs such as the wood willowherb (*Epilobium angustifolium*) or foxgloves (*Digitalis* spec.), which are absent in permanent forests, settle here. However, neophytes such as Indian touch-me-not (*Impatiens glandulifera*) and giant hogweed are also spreading. For reptiles such as the sand lizard or the adder (*Vipera berus*), too, the forest often only becomes habitable through clear-cutting.

However, both old-growth forests and permanent forests have the major disadvantage in common that old trees with old and dead wood are almost completely absent in them, indeed are not even provided for in the forestry management concept in the first place. Especially old wood from dying trees and dead wood from dead trees is vital for numerous insects (Insecta) and their larvae as well as for many species of fungi (see Chap. 21). To achieve this, however, forest areas must be partially or completely taken out of use, as envisaged in the concept of a natural (protected) forest. This has already been addressed elsewhere (Chap. 14, Section "Forest Ecosystems") (Figs. 24.2 and 24.3).

Fig. 24.2 Merian engraving of Heidelberg (image source: Immanuel Giel, Wikipedia) from the early seventeenth century. The illustration shows that the deciduous forest only covers the areas immediately adjacent to the city as well as the valley hollows of the Odenwald (with deeper soil formation) and probably serves as a winter firewood resource for the city. In contrast, the heights around the Gaisberg (lower and to the right in the picture) as well as the Königstuhl (higher and to the left) are free of forest as a result of extensive grazing and, in the absence of fertilization, can be interpreted as rough grassland. Under the base-poor conditions of the Buntsandstein, this is formed as a borstgrass lawn with broom heath

Fig. 24.3 Photomontage (with shortening of the intermediate section) of approximately the same section as in the picture above (the right summit of the Gaisberg, actually lower, appears higher in perspective due to its proximity than the Königsstuhl in the picture on the left; this was "corrected" in Merian). The advance of the forest onto the ridge in the meanwhile past four hundred years becomes clear: essentially a consequence of forest management since the eighteenth and nineteenth centuries. (Own recording)

25

Agricultural Accompanying Programmes of Nature Conservation in Germany

Field Margin and Flowering Strip Programmes

For at least half a century, the originally numerous species of field **weeds** (Fig. 25.1) have largely disappeared from the fields, including beautifully flowering and widely appreciated plants such as corn poppy (*Papaver rhoeas*), cornflower (*Cyanus segetum*), corn cockle, field delphinium (*Consolida regalis*), camomile or field pansy (*Viola arvensis*). Many of these field wild herbs were almost perfectly adapted to life in the field, including the fact that they matured with the grain and the seeds were thus inevitably sown again and again with the retained grain seed.

However, **seed cleaning** and **herbicides** have done a good job in the meantime. Some of the field weeds have not become rare because they found similar, short-lived sites with unhindered light incidence on ruderal areas. Among the species mentioned, this is widely true for the bright red corn

K.-D. Hupke, *Nature Conservation*,
https://doi.org/10.1007/978-3-662-66159-8_25

Fig. 25.1 Traditional field weeds such as chamomile (*Matricaria chamomilla*), creeping thistle (*Cirsium arvense*) and corn poppy (*Papaver rhoeas*) (all pictured) have largely disappeared from cereal fields thanks to seed cleaning and herbicides. Field margins can only provide partial relief. (Own recording)

poppy and for the true chamomile (*Matricaria chamomilla*). Other species such as the corn cockle (*Agrostemma githago*), the field spadefoot (*Thymelaea passerina*) or the Venus mirror (*Legousia speculum-veneris*) became largely extinct.

However, the frequently expressed view that an increased expansion of field **margins**, e.g. between different land use and ownership parcels or along paths, could provide a refuge for the endangered field wild herbs must be contradicted. Basically, a field that is either fallow in spring or sparsely overgrown with winter cereals is a different growing habitat than a marginal strip. Even if the latter is mown once a year or more frequently, it is more like a meadow than a traditional cereal field in terms of the density of vegetation cover and precisely because of this mowing. In addition, as a narrow marginal strip, it is subject to a greater input of fertilizers and pesticides from the field, so that its naturalness must be strongly questioned. At least it provides a fodder base for many wild animals such as partridges and field hares with a stronger growth of wild plants and also offers a substitute for the increasingly declining grassland, especially at lower altitudes.

Field margin programmes are funded by the European Union and also by many German state governments with subsidies. The farmer is allowed to use the field margins in the course of these programmes by mowing, but is subject to restrictions there, especially in the use of agrochemicals.

Strictly separate from field margins are so-called **flowering strips** (narrower) or **flowering fallows** (large areas; Fig. 25.2), which are not colonized autonomously, but where flowering seed mixtures of almost exclusively

Fig. 25.2 Flowering fallow in the northern Alpine foothills near Augsburg; late summer aspect. Preferably beautiful flowering ornamental plants are cultivated. Their planting is controversial in nature conservation. Especially the sunflower seeds are accepted by many field bird species in the winter half-year. Biodiversity is increased in some respects compared to the rest of the intensively used agricultural land. However, this is not "nature" in the sense of nature conservation. A suggestion here would certainly be to favour site-specific seeding by the subsidising legislator. (Own recording)

non-native species are sown. Within the framework of the EU **Greening Programme**, there are considerable subsidies for the establishment of flowering strips, which amounted to about 600€ per hectare in 2017 (in comparison: in the same calendar year, wheat cultivation yielded an economic return of about 1000€ per hectare in the northern Upper Rhine region; however, with increased labour costs for the farmer for pest control and harvesting).

Field margins, flowering strips and flowering fallows are playing an increasing role as fields of action against **insect mortality**, which has also been recognised by the mass media since 2016. The partially alarming results do not yet seem to be sufficiently verified by spatially broadly based long-term studies, which will not be achieved so quickly due to the time factor. The causes of the assumed insect mortality, i.e. decline in both mass and number of species, are suspected to lie in different areas. On the one hand, the use of pesticides, especially neonicotinoids, is discussed as a cause. On the other hand, the intensification and standardisation of the agricultural landscape allows fewer and fewer refuges both within the cultivated areas, where glyphosate radically reduces all unintended wild plants ("field weeds"), and at the edges in the form of field hedges or field margins. Field margins, flowering strips and flowering fallows are intended to counteract this and provide both feeding areas for beetle larvae and butterfly caterpillars, for example, and nectar sources for (wild) bees and butterflies, among others.

Set-Aside and Extensification Programmes

State programmes of the EU, the federal government and the state governments for set-aside and land extensification are integrated into the **Common Agricultural Policy**

(CAP) of the European Union. After the Second World War, this policy was initially characterised by the attempt to increase land yields. This was an urgent requirement in the immediate post-war period in order to ensure food security for the population.

This policy of **increasing food production** guaranteed farmers high producer prices on numerous products such as wheat, milk and beef, completely independent of the respective world market prices. The surpluses, which inevitably grew with increasing yields per hectare, went to state warehouses and produced steadily growing "grain mountains", "butter mountains" and "milk lakes" until the 1970s. These state stocks of food could often only be reduced by exporting the corresponding products to the world market. This could only work because these agricultural products were drastically reduced in price for export so that they corresponded to the price level on the world market. This regularly caused scandals in the media when, for example, butter went to "the Russians" cheaper than German consumers could buy it.

The general unease with this traditional agricultural policy led to a complete change in agricultural support in the 1980s. In future, subsidies to farmers, at times more than half of the EU budget, were no longer to lead to an increase in agricultural products. At least some of the subsidies were paid for the sheer opposite: for simply leaving agricultural land lying idle and not cultivating it. **Set-aside premiums** were paid to farmers. Perhaps farmers never had an easier time getting their income than they did at this particular time (not to mention the bureaucracy involved in applying).

Land extensification programmes such as the Baden-Württemberg MEKA (Marktentlastung und Kulturlandschaftsausgleich – market relief and compensation for cultivated land) go in a similar direction. In particular, it promoted the abandonment of agricultural chemicals (synthetic fertilizers, pesticides). The advantage over set-aside programmes was twofold: on the one hand, the corresponding areas, largely grassland, continued to be mown regularly and therefore did not appear "overgrown", unlike fallow land. On the other hand, no ruderal plants spread on the respective areas, but rather the traditional meadow plants, which are more highly valued by nature conservation and which prefer precisely such nutrient-poor habitats: for example arnica (*Arnica montana*), numerous species of bellflower (*Campanula* spec.), orchid species. Thus, a piece of traditional agricultural landscape has been and is being restored.

However, land extensification as well as **set-aside** had the same effect. In the course of the 1980s, all agricultural "mountains" and "lakes" declined drastically and EU food production gradually approached EU food consumption. EU-produced food was still too expensive for large-scale export to the world market. The scheme thus satisfied farmers, public authorities, nationals and competing agricultural exporters on the world market alike.

However, the set-aside programmes have also largely expired or have not been renewed. The reason for this is that, at the beginning of the new millennium, the world agricultural market, and with it the EU's agricultural situation, underwent another fundamental change. As a result of the rise in the price of fossil fuels and laws in probably all industrialised countries promoting greater use of renewable energies, there has been an increase in the use of agricultural products for energy production (biodiesel, ethanol, methane). The traditional task of agricultural land to produce food thus came into ever stronger competition with the new task of producing energy. As a result, agricultural products such as farmland are becoming increasingly scarce and expensive. Agriculture, which in economic terms has always lagged behind manufacturing and the service sector in terms of growth prospects, now seems to be heading for a golden future. From the point of view of many, however, set-aside and extensification of land, the aim of which was to limit agricultural yields, have clearly come to an end.

Meadow Breeding Programs

Most bird species are known to cavity breed or breed in shrubs or trees. A smaller number are ground breeders, with some species preferring forests, others fields. Some species also lay their nests in grassland, known as **meadow-nesting birds**. Perhaps the best known meadow-nesting birds in our avifauna are the skylark (*Alauda arvensis*) and the lapwing (*Vanellus vanellus*). *In* addition to their preferred grassland, both also breed on arable land.

Although the skylark, and to a lesser extent the lapwing, can still be considered common in some regions, their populations have declined. This is due to the increasing intensification of agriculture. The preference for winter cereals and the considerable fertilisation of meadows and cereal fields result in high plant growth, especially in spring, which keeps the nests of the lapwing and the lark, both of which naturally come from dry steppe areas, strongly shaded and moist. The population development of the lapwing shows that especially years with a cool and wet spring lead to population declines if the broods cannot be reared.

Another point of danger for meadow nesting birds is that the first mowing takes place so early due to the increased application of fertilizer that the brood has not yet fledged and falls victim to the mower.

Meadow breeding programmes, as provided for in the Federal Nature Conservation Act and supported by most Länder with their own programmes, are intended to serve the extensification of grassland. In addition, the increasing conversion of grassland into arable land is to be counteracted (so-called EU greening programme from 2013). Financing is provided within the framework of so-called cross-compliance (meaning mutual obligation; here the state and the farmer are meant) from the public agricultural budgets, which have been largely decoupled from the level of agricultural yields in the course of a reorientation of EU agricultural policy in recent decades, in the form of premium payments to farmers who have to comply with certain nature conservation and environmental protection requirements.

26

Europe is also Getting Involved: German Federal Conservation Act, FFH and Natura 2000

The **Reich Nature Conservation Act** of 1935 remained valid in principle as a framework law long after the Second World War, now as the **Federal Nature Conservation Act**. It was only in 1976 and 2002, respectively, that it underwent a thorough revision by social democratic governments. The background to the new regulations was a growing awareness of the finite nature of natural resources, as reflected in the pronouncements of the Club of Rome, for example, or the new term "environmental protection" (around 1970). The Reich Nature Conservation Act, which was considered progressive at the time, was less and less able to meet these comprehensive requirements. In addition, the terminology and content of nature conservation were increasingly determined by an international discourse.

However, over the past 20 years, policy and legislation have also shown an overall increase in the **influence of the EU** on the situation in the national Member States beyond the framework of nature conservation. This is no different

K.-D. Hupke, *Nature Conservation*,
https://doi.org/10.1007/978-3-662-66159-8_26

in nature conservation and also seems obvious given the situation that natural areas mostly occur across borders and animal migrations do not stop at borders. EU law usually provides a narrower or broader framework within which national legislation can develop. In addition, cross-border programmes are also coordinated from Brussels.

The need for large-scale international nature conservation regulation is particularly obvious in the **protection of migratory birds**. Accordingly, the earliest and most comprehensive European regulations are found in this area. As early as 1979, a joint directive of the then EEC on the conservation of wild birds (later part of Natura 2000) was adopted, which had to be directly transposed into national law. This resulted in the designation of almost innumerable **bird protection areas** in Germany and Austria. However, there was already an international model for this: the Ramsar Convention of UNESCO for the protection of waterbird habitats from 1971.

The European **Natura 2000** concept was and is closely linked to the fauna-flora-habitat **concept (FFH). In** accordance with the global *Convention on Biological Diversity*, Conference of Rio de Janeiro 1992, the aim is to halt the ongoing loss of biological diversity. To this end, around 20% of the national territory has been designated as FFH areas throughout Europe to date. With such large areas, it is also clear that nature conservation will not be the top priority everywhere and that compromises with agriculture and forestry are unavoidable. The required areas are declared in close cooperation between the European Commission and the member states. In the German national legal context, these areas are known as nature and landscape conservation areas; in Austria, they are mainly included in the so-called *European protected areas* created specifically for this purpose. The already diverse status forms of protected areas in the national context are further complicated by the EU terminology, so that one and the same protected area often has several protection terms.

In the areas of agriculture and food as well as nature conservation and environmental protection, Europe has developed the most towards a political union of all social areas; also with the result that we have to live more according to central guidelines, whereas nature conservation in Germany, as in Austria, is traditionally primarily a matter for the federal states.

In addition to the EU, global nature conservation projects are increasingly having a regulating effect on national legislation, even though, unlike the EU, they are mostly not legally binding. This applies in particular to the complex surrounding the UN Conference on Environment and Development (UNCED) in Rio in 1992 with the **Convention on Biological Diversity (Biodiversity Convention)** and the international **Washington Convention on** International Trade in Endangered **Species of Wild Fauna** and Flora **(CITES)** in 1973, which regulates trade in endangered animal and plant species. Germany and Austria are also involved in the international UNESCO concepts of **biosphere reserves** and **world natural heritage**. While the Rio process (including the Rio follow-up conferences) and the concept of biosphere reserves strive for a balance between nature conservation concerns, such as biodiversity, and aspects of economic use under the aspect of *sustainability*, the world (natural) heritage primarily serves educational aspects. However, both concerns are traditionally already integrated in nature conservation. The aspect of sustainability in particular has become very complex in recent decades and would require a very comprehensive monograph of its own. Since the diversity of its aspects far exceeds that of nature conservation and, moreover, is more closely related to environmental protection, it will be largely omitted here.

27

On the Role of Non-Governmental Organisations (NGOs) in Nature Conservation

As is probably the case with all social objectives, nature conservation did not emerge from the outset as a governmental project at the level of the authorities, so to speak. In the beginning, both in Germany and, as will be shown in Chap. 30, in North America, there was a need to preserve what was threatened. Through the establishment of corresponding private institutions that acted as pressure groups vis-à-vis the media and society, it was also possible to anchor the concern institutionally via politics, for example by drafting state laws and setting up specific authorities within their framework.

It is therefore not surprising that from the beginnings of European and North American nature conservation in the nineteenth century to the present day, non-governmental organizations (**NGOs**) have played and continue to play a central role in nature conservation. Even if today the main financial support comes from government agencies, it remains an important function of nature conservation

K.-D. Hupke, *Nature Conservation*,
https://doi.org/10.1007/978-3-662-66159-8_27

associations to ensure that public and media pressure on governments and parliaments does not diminish to do more for nature conservation.

For the reasons already explained in Chap. 3, however, nature conservation associations in the narrower sense, which in this country are united in the **Deutscher Naturschutzring**, have lost relative importance in recent decades compared to environmental protection associations such as Greenpeace or animal protection associations such as PETA. Relative does not mean absolute, however, as the influence of NGOs on society and the media has increased almost continuously in recent decades.

The most important nature conservation associations in Germany are the **NABU** (Naturschutzbund Deutschland; in Bavaria: LBV) and the **BUND** (Bund für Umwelt und Naturschutz Deutschland; in Bavaria: BUND Naturschutz); in the south-eastern neighbouring country the **Naturschutzbund Österreich**. All three have been able to defy the loss of importance of nature conservation concerns since the times of the "television professor" Grzimek in the 1960s – at the price of a partial adaptation of their goals to environmental protection. In the case of BUND, this is even given equal status with nature conservation in the name. In the case of Naturschutzbund Österreich, the link to environmental priorities is the least pronounced in comparison.

However, such an adaptation is in principle not possible without abandoning part of the nature conservation objective: Today, all the nature conservation associations mentioned above advocate the principle of renewable energy resources, which, through price increases for agricultural products, leads to further agricultural intensification, and in developing and newly industrializing countries also to a further expansion of agricultural areas at the expense of near-natural reserve areas (as well as, in competition with the production of food, to an intensification of hunger).

NGOs are in any case extremely dependent on the media and public opinion. This dependence also exists in the opposite sense, since NGOs today very strongly shape public perception, especially since they are granted that they work more disinterestedly and honestly than the two other social subsystems of economy and politics.

NGOs therefore have to market themselves. Since the pure topic of nature conservation is often abstract, they have to resort to symbolic animals in the form of mascots in order to become effective in the media. For the international **WWF** (World Wide Fund for Nature), this has for decades been the panda, occasionally assisted by big cats such as the tiger or jaguar and by apes such as the orangutan. And, at the latest since the hype about Knut, of course the polar bear (*Ursus maritimus*).

Just how high the affinity to animal protection is, especially in the WWF, which is committed solely to nature conservation concerns by virtue of its statutes, was demonstrated in the summer of 2012 when King Juan Carlos had to resign as Honorary President of WWF Spain. He had hunted elephants in southern Africa and had his picture taken in front of a shot elephant. Unlike in sub-Saharan and eastern Africa, elephants (Loxodonta africana) are not threatened in the south of the continent. As they have hardly any natural enemies other than humans, they tend to breed en masse and destroy entire forest areas, woody plants and numerous individual trees in a savannah landscape characterized by small-scale alternation of tree stands and grassland (Chap. 30).

Of course, not only NGOs need a good **external image**, but also the companies of the economy. Thus, there is regularly an alliance between companies and NGOs, with the former buying a good image with monetary gifts. Especially the WWF ("Partner of the Economy") has become known, some would say infamous, through such close **cooperations**. The WWF is therefore probably by far the most financially successful international nature conservation organisation, with an annual budget of several hundred million euros. It goes without saying that such flows of money also create a dependency of the NGO in question on the donor, who can buy his way out of responsibility with relatively little money and also use the WWF logo for his advertising.

Unlike most other conservation organizations, WWF also has a thriving youth organization, WWF Young Panda. Together with the business model, this should also help WWF remain the most influential conservation organization, at least in the global arena.

28

The Silverware of the GDR? Nature Conservation and Conservationists in the Eastern Federal States

While many once common animal species had been eradicated or virtually extinct in the western federal states by the time of reunification, they were often still to be found in surprisingly large populations in the GDR. This applies, for example, to the large birds of prey Osprey (*Pandion haliaetus*) and White-tailed Eagle (*Haliaeetus albicilla*). It is true for cranes. And it is true for beavers and otters. In the Federal Republic of Germany, a large proportion of these species were not subject to nature conservation law, but to **hunting law**. This, however, was shaped by a significant lobby of hobby hunters, which included many big names from finance and industry as well as politicians such as Franz Josef Strauß, against whose resistance it was hardly possible to push through a law over a long period of time. Thus it took until the 1970s until a general protection for the birds of prey came about. By then, however, ospreys and white-tailed eagles had been all but wiped out in the Federal Republic. The recovery of the populations of most of the

K.-D. Hupke, *Nature Conservation*,
https://doi.org/10.1007/978-3-662-66159-8_28

species mentioned from then on in West Germany, which often resulted from immigration from the East, was due, among other things, to the fact that the influence of hunters on West German politics was gradually reduced, not least due to the death of Strauß (on a hunting trip, by the way).

That hunters kill animals out of their self-image is self-evident and part of the definition. But why do hunters kill animals to the point of complete extinction, thus depriving themselves of the possibilities of their hobby? To understand this, one must first consider the smallness of most hunting grounds, and that a hunter who spares an animal because of its rarity must expect that the owner of the neighboring hunting ground will kill it. On the other hand, hunters are also trophy collectors. The rarer a species, the more spectacular the trophy.

Especially the birds of prey were the special dislike of many hunters. They were thought to strike young hares and partridges. They were thus a direct "food competition" for the hunters.

Most of the species mentioned require diverse, but not necessarily use-free areas and often fit in well with agriculture and forestry. In this respect, there was not necessarily a contradiction between use and protection in the GDR, just as the tendency to maximise agricultural profits initially emanated more from the West than from the East (although the East often followed these Western developments, if only to be able to survive in the competition between systems).

The Grand Canyon and the Amazon, the Himalayas and the Maldives, but also the Côte d'Azur and Tuscany remained largely closed to almost all GDR citizens interested in nature. This explains the greater devotion of many people to their own, nearby nature. Similar to Western societies, the political reality of the GDR encouraged people to enter worlds of escape. Alongside artistic fiction, nature formed the best and literally closest possibility of escaping social reality.

Compared to Western societies, **nature conservation in the GDR** was less characterized on the side of the actors by independent nature conservation associations, which would probably not have been tolerated as competition to the party and its sub-organizations. Individuals were more active than institutions in the GDR—and still are in the eastern federal states. Occasionally, these individuals became legends in old age or after their demise. Kurt Kretschmann (deceased 2007), honorary president of NABU after reunification, is one of them; Michael Succow (born 1941) is another.

Through skilful interaction between the already freely elected transitional government under Hans Modrow and the then West German Environment Minister Klaus Töpfer, it was possible to transfer a considerable proportion of the remaining large areas close to nature in the former GDR to the status of **nature reserves,** if this had not already existed. Shortly thereafter, many of them became **national parks in** the Federal Republic of Germany, which in their self-image followed the US model and which therefore were not allowed to exist in this designation in the GDR.

For many people in the new Länder, the continued existence of nature saved an essential part of their identity from the GDR to the new Federal Republic. As in parts of the welfare state (e.g. day nurseries as a regular service), the new federal states thus became pioneers for a (gentle) structural reform of the old Federal Republic as well.

As a result of this historical development, nature conservation in Germany today is, on the whole, more strongly influenced by the eastern federal states than would correspond to their share of the total German population. At the same time, East German nature conservation has retained its own characteristics. As a recent documentation of renowned East German conservationists in the work *Naturschutz in Deutschland* i.e. Nature conservation in Germany (Succow et al. 2012) shows, East German conservationists never really arrived in the West-oriented reunified Germany. "Nature conservation in East Germany with some additional information about the West" would actually have been the appropriate title for this book. Yet, despite the average retirement age of its actors, East German nature conservation seems surprisingly vital and, above all, authentic. Even after reunification, East German nature conservation has never made the many advertising compromises with regard to the self-presentation and external image of an NGO, as was the case with the (West) German nature conservation of the past decades, including relevant goals and aspects of environmental protection and animal welfare, which even its own clientele is often unable to distinguish clearly. But a distinction between these conflicting objectives was and is not wanted in the West, because then one would have to reveal the conflict and often incompatibility of the diverging objectives. In the process, however, the goals of nature conservation have somewhat dissolved their contours. In comparison, East German conservationists have remained such in the literal sense of the word.

In the GDR, too, the contemplation of nature and efforts to protect it provided an escape route from social reality, even though, or precisely because, very few East German conservationists were explicitly critical of the GDR during its lifetime. Almost more so than in the West, the nature conservation clientele in the East is outdated in terms of personnel; not least because nature conservation there has made so few promotional and fashionable compromises that would have attracted young people to it. Paradoxically, however, nature conservation seems to be particularly successful as a spatial planning approach where it makes few compromises in terms of content and leans less on the zeitgeist.

No question, I am sympathetic to the conservationists in the East. Nevertheless, I have devoted only a small chapter to them because the main tendency of the development after 1990, as in all political, economic and social issues, emanated from the former political West of Germany and still does today. In addition to the demographic preponderance of West Germany, this is also related to the fact that the 1989/1990 merger was the victory of one social system, which now codified its structures, over another.

Nature Conservationists of the Former GDR and the Subsequent Period of Political Change About Their Nature Conservation Opportunities at That Time

Species that were less in the spotlight were also worked on. For example, we had colleagues who were very strong advocates for gentian species, for sundew, or for arnica. We made assessments of population trends in the selected plants and animals as well as very clear recommendations for biotope protection and biotope maintenance. These suggestions and indications were taken up. In this respect, one can already say that it was a very productive time in which something was achieved in nature conservation.

(...)

We didn't complain about money for species conservation. We did not have large sums of money at that time. But the research we did with the members of the interdistrict working group on species protection in the nature reserves was financed, among other things, by the district council in Suhl. For example, when I wrote a letter to the factory manager of the "Stern-Radio" factory in Sonneberg and told him that I needed four people at a certain time for investigations in a nature reserve, he provided me with four employees from the factory. Such measures took place at that time in the context of socially recognized work. The people concerned were paid their wages or salary normally for the time. Everyone accepted this. Necessary travel expenses were reimbursed by the district council. That would be unthinkable today!

(*Martin Görner, born 1943; nature conservation officer of the district of Jena-Stadt and the district of Gera for many years*)

During the Wende and after I returned to Schleiz, there was a euphoria. There was excellent staffing that nature conservation had in the county. Agriculture was open to different things. There were new forces to work with. I was and am a volunteer conservation officer. The forest conversion program in forestry was excellent. I have to tell you: it was a relatively short-lived joy. It then developed into something where everything was sacrificed to money. Either you have money or if you want to make a lot of money, you have to work against nature in this or that form. To make a long story short: The euphoria has given way to disillusionment. And we have to fight for nature conservation now just as we did in GDR times.

(*Günther Hoffmann, born 1935; among other things, former nature conservation officer in the former East German and later all-German Ministry of Agriculture*)

It was a wonderful time, we had survived the war, life lay ahead of us! The surroundings of Eberswalde enticed us to explore, with school friends I roamed through forests and moors and also discovered the oldest nature reserve "Plagefenn", which could be saved from axe and saw since 1907. I remember a hike with schoolmates to the Plagefenn. We stood reverently in front of the famous "Drehkiefer", that strong tree that had been photographed lying on the ground in the 1930s and was still there now, in 1950. Little changed, I photographed this pine myself in 1972 and even today the knowledgeable can still find it! In general, the Plagefenn, it has decisively shaped my idea of what nature conservation is or should be.

(*Lebrecht Jeschke, born 1933; employee of the Institute for Regional Research and Nature Conservation in Halle, after reunification Nature Conservation Academy Vilm and Head of the National Park Authority Mecklenburg-Western Pomerania*)

The opportunities for forestry to make an active contribution to nature conservation have also changed; the privatisation of parts of the forests and also in the state forest are managed with a very high focus on yield. Before reunification, forestry operations still had some leeway, for example a budget for rural cultural measures that was not at all small and opened up opportunities for conservation measures. Through commercialisation, the connection between forestry and nature conservation has loosened considerably.

(*Heinz Quitt; born 1928; senior forester and forestry office manager, nature conservation officer of the Magdeburg district*)

When I then travelled to the "West" in the summer of 1990 and experienced industrial agricultural production in Höxter and Diepholz, I was shocked at what continued to happen there "with impunity" in the countryside. We had first largely overcome factory farming with the fall of the Berlin Wall, and now in Lower Saxony (with well-kept villages, a church in every village, devout people) I was able to experience industrial agricultural production with feed imports from overseas at an intensity that exceeded much in the GDR.

(*Michael Succow, born 1941; qualified botanist, after reunification Deputy Minister for Environmental Protection of the GDR, Professor at Greifswald University; winner of the Alternative Nobel Prize*)

From: Behrens and Hoffmann (2013)

29

The World's Oceans and Antarctica: International, Therefore Defenceless?

In a global view, we must always keep in mind that only part of the earth's surface is divided territorially among nations. For the continental area, such a division applies almost universally. Nevertheless, Antarctica is also excluded here; at least national claims to ownership there are not internationally recognized.

The situation in **Antarctica** in the post-war period was (and still is) characterized by the fact that the world powers USA and Soviet Union (today: Russia) did not recognize any claims of ownership by individual states. Therefore, instead of a territorial division favoured by the riparian states, but also by Great Britain, France and Norway, a treaty was established that regulated the possibilities of using Antarctica or the continent of Antarctica. The so-called Antarctic Treaty was adopted in 1961. It opens Antarctica to all interested nations for peaceful research and excludes military use.

In the 1980s, a campaign supported by Greenpeace for a "**World Antarctic Park**" prepared the "Madrid Protocol"

K.-D. Hupke, *Nature Conservation*,
https://doi.org/10.1007/978-3-662-66159-8_29

of 1991 as an additional agreement to the Antarctic Treaty. The Protocol bans resource extraction for the next 50 years, starting in 1998 (the year it came into force, after enough countries had signed it). The protocol also imposes environmental requirements on research stations. What happens after 2048 (end of treaty, extension or renegotiation?) is still open. Greenpeace also complains that tourism, which is growing rapidly in the marginal area of Antarctica, is not yet subject to any contractual regulation.

Nature conservation and environmental protection are promoted in the area of Antarctica by the fact that, for cost reasons, commercial extraction of mineral resources on the Antarctic continent, as well as in the surrounding shelf areas and deep-sea regions, appears to be ruled out for the time being. The question is how the Parties will proceed once this circumstance has changed, for example through technical progress.

In the case of the **world's oceans**, territorial allocation was limited to a narrow coastal strip of three nautical miles until the 1970s. In the meantime, the national sovereignty zone has been extended to twelve nautical miles and is supplemented and expanded by an **exclusive use zone** of 200 nautical miles. Here, the riparian states have exclusive economic exploitation rights, for example to fishing or to mineral resources on the seabed.

Outside the exclusive use zone, which according to WBGU (ed. 2013) still covers around 64% of the world's oceans, no use requirements are traditionally envisaged. An agreement under the UN organization UNCLOS (United Nations Convention on the Law of the Seas) is emerging; however, all agreements remain non-binding in principle as long as the USA has not signed the agreement.

It is precisely this zone, which has not yet been allocated nationally but is predominant in terms of area, that can be considered particularly endangered because restrictions on fishing do not apply there. Most **marine mammals** and **sea turtles** (Cheloniidae) are protected by international species protection agreements (CITES), but often fall victim to the extensive trawling and drift nets as bycatch. For most marine cetaceans (as a result of the oldest international conservation agreements for the seas, some dating back to before the Second World War), stabilisation at a low level has at least been achieved, although population trends vary greatly depending on the species. In contrast, the populations of the six species of sea turtles (Fig. 29.1) are still tending to decline.

After all, the expansion of exclusive economic zones has meant that the vast majority of **coral reefs have** come under national sovereignty, as have the **mangrove zones** (Fig. 29.2) of the expanded estuaries. These two such contrasting habitats (clear water/muddy) harbor a great wealth of animal species, which are also largely known to us. It is

Fig. 29.1 Walled tank of a sea turtle (Cheloniidae) "hatchery" in Sri Lanka. The animals caught by the fishermen are displayed to tourists for an entrance fee. A far greater danger for sea turtles, however, is that they often get caught in oversized fishing nets and subsequently drown. (Own recording)

Fig. 29.2 Mangrove trees (as seen here in South Florida) usually form stilt roots, among which a variety of fish and crab species live as the water level changes. (Own recording)

possible that the **seafloors of the deep sea** below the international high seas are hardly inferior to these habitats, at least regionally, in terms of biodiversity. However, knowledge about these zones is limited, as is the presumed human influence on them.

Since coral reefs are mostly under national sovereignty, they can also be designated as nature reserves. This has happened in isolated cases, for example in parts of the Great Barrier Reef, the largest contiguous coral reef on earth, in north-eastern Australia (Queensland). However, this hardly curbs spatially overlapping dangers: for some time now, damage to corals due to ocean warming and excessive nutrient inputs from wastewater and agriculture has been discussed.

But here, too, the search for causes is proving to be extremely complex. The strong tendency of corals to die off in the past decades (coral bleaching) was partially replaced by a recovery phase. Possibly a natural life cycle of the reefs is also involved. Diving and snorkelling tourism, on the other hand, must almost be taken in its stride. It does lead to the local death of reefs, especially when the sensitive coral surfaces are stepped on, which is always deeply shocking at first sight. However, in all the main tourist regions only a few percent of a large reef is affected. Away from the hotel complexes, these influences hardly play a role. Moreover, hoteliers and other groups profiting from tourism have a strong interest in ensuring that the corals in front of the hotel beach do not die and are increasingly counteracting this with educational work (Fig. 29.3).

Today's increased technical possibilities and economic needs with regard to the use of marine and (anti)arctic resources are matched by an increasingly greater effectiveness of international restrictions and regulations on use. However, it appears that protection regularly lags behind expanded use.

Fig. 29.3 Fish of a coral reef near a hotel are fed by tourists from a boat and gather at the water surface (Hikkaduwa, Sri Lanka). (Own recording)

EU Agrees on Fish Catches

The EU states are doing more to protect fish stocks than in the past, according to environmentalists. Many North Sea stocks have come out of the worst years and are now at safe levels," said Stella Nemecky, referring to the higher quota for plaice (*Pleuronectes platessa*).

[...]

The EU wants to manage its fish stocks, some of which are under strain, more carefully in future. The framework for this is provided by the fisheries reform adopted in the spring. The State Secretary in the Federal Ministry of Agriculture, Robert Kloos, explained: "More than 60% of stocks in the North Sea and North-East Atlantic are now managed sustainably. By 2015, this percentage will continue to increase significantly, and by 2020 at the latest, we want to reach the sustainability target for all stocks."

Environmentalists, on the other hand, point to about 40% of the stocks that continue to be overfished, "The overfished herring stock in the Kattegat and Skagerrag may be fished about three times more than scientists had recommended. The active fisheries for cod and sole (*Solea solea*) in the Irish Sea were not closed, contrary to scientific advice, the WWF complained" (*Stuttgarter Zeitung* of 19.12.2013, according to a report by dpa).

Antarctica is Now Protected

The largest marine protected area in the world is to be created off the coast of Antarctica: The protected area in the ecologically important Ross Sea is to cover 1.55 million square kilometres. This is provided for in an agreement reached by 24 countries and the EU after years of negotiations in the Australian city of Hobart.

Environmentalists speak of a "historic decision", but criticise that the agreement is only valid for 35 years. The designated zone is about the size of Britain, Germany and France combined. In the largest part of the new protected area—1.12 million square kilometres – all fishing is to be banned.

Report AFP n. Stuttgarter Zeitung v. 29.10.2016.

Indigenous Fishermen Fear Deep Sea Mining

The inhabitants of the north-east coast of Papua New Guinea and the offshore islands of New Britain and New Ireland are following with concern how deep-sea mining is being prepared in the Bismarck Sea. Canadian company Nautilus Minerals Inc. plans to launch the world's largest deep-sea mining project to date, Solwara 1, in the sea area, which is more than 1000 m deep. Starting in 2019, the company plans to explore and mine raw material deposits on an area of 500,000 km^2 [far more than the size of Germany; d.Vf.]. Discoveries of valuable minerals and metals such as gold or copper are expected. The company has already secured a mining licence for the next 20 years from the Papua New Guinea government. But among the indigenous Melanesian inhabitants of the islands, as well as many non-governmental organizations, skepticism and rejection prevail.

The Papuan people who live from fishing fear the ecological consequences of Solwara 1, because no one has ever had experience with such a mining project at such a great depth. The fishermen fear that in the long term the abundance of fish could decline. The churning up of millions of tons of debris in the deep sea could destroy spawning grounds and alter fish migration routes. It could also affect coral reefs, which are vital to marine life and fish abundance. Critics accuse the government and the mining company of not having sufficiently investigated the possible consequences and of not being transparent about the circumstances surrounding the granting of the mining license.

(Ulrich Delius 2016, p. 46).

30

How Is Nature Protected Outside Europe?

The Example of the USA

As a teenager, I struggled through the monumental work *Leatherstocking* by James Fenimore Cooper, and I still remember one scene: the already old trapper leans on his shotgun and watches uncomprehendingly the goings-on of his contemporaries, who gun down a herd of bison in a kind of bloodlust without having any direct use for meat or leather. The scene marks a change in intellectual history. It becomes increasingly apparent that the nature of the new continent is not as inexhaustible as the early colonists had assumed and as is still ingrained in many minds today. After the migratory pigeon had already been exterminated, it was now the turn of the bison; both species of animal that had been widespread on a massive scale in pre-European times and were in principle considered inexhaustible. The life of the trapper Leatherstocking, whose real name was Nathaniel Bumppo, turned out to be a flight away from advancing civilization.

K.-D. Hupke, *Nature Conservation*,
https://doi.org/10.1007/978-3-662-66159-8_30

Consequently, the protagonist had also spent young adulthood in the forests of Appalachia in the East and his life ended on the prairies in the West. Had Leatherstocking, improbably, lived on for a few more decades, he would probably have faced real problems, since even further west are the semi-deserts, which really can't provide an existence for a trapper and hunter. But Leatherstocking had always been one, who killed only as much as he needed to live. One who lived like the "wild" Indians, with whom he was also friends as long as they still existed, and whom he basically understood better than his white compatriots.

Even though the author Cooper's criticism of the immoderate settlers is clearly formulated in the work: It is a question whether these trailing millions as trappers and hunters would not have led to a still more rapid ruin of nature. The settlers were primarily arable farmers. This means that, compared to the hunters, they were able to feed a family on a relatively small area of land. Above all, the large mass of European settlers made them a problem compared to the relatively few Indians and white trappers before.

When settlers arrived at the arid agricultural frontier in the West in the mid-nineteenth century, it reminded Americans for the first time in their history of the finite nature of their resources. A little later, and probably not entirely coincidentally in this sequence, the first large conservation areas were designated in the form of **national parks**. The immigrants now arriving in ever greater numbers in the second half of the nineteenth century had to stay in the coastal cities of the East (or the far West in California) and find work there in the rapidly emerging industry and growing cities. After unlimited opportunities in agriculture now came an unlimited rise in industry, which by 1890 had already elevated the **United States to the** greatest economic power on earth and what would later become a superpower.

Despite, or perhaps precisely because of, their intrinsically unnatural living environment, this growing urban population idealized a nature from which they did not directly live, but which they wanted to preserve. This distant nature of the national parks could not and was not allowed to be anything other than grand and unlimited, thus American. At the attached campgrounds, life on the Frontier Line could be re-enacted and relived in short vacations, albeit as a limited experiment.

However, the gigantic nature that had become tangible in the national parks, especially in the US West, also closed a gaping wound in the self-esteem of the still young nation. Almost every educated European traveler who visited the United States in the early nineteenth century noted the glaring lack of historically-grown culture. Now this deficit was not something that could be changed anytime soon in the process of nation building. And yet the United States seemed to have come away with something that seemed at least as valuable as the remnants of a millennia-old European culture. There were waterfalls and canyons higher and deeper, respectively, than the tallest European cathedrals. There were trees older than European cultural history. And the creators of this natural wealth were not philosophers, architects, and artists, but God's own hand had created these riches of nature and given them to the United States, his obvious favorite nation, to enjoy and edify. It was now the task of American society to preserve this gift from destruction and to perpetuate the ongoing enjoyment of it. National parks were thus territorial exceptions within an economic mindset that viewed available land almost exclusively as economic capital and a factor of production. However, they were and still are also a distinct social corrective. In this way, the natural spectacles of the national parks became important symbols of national self-image.

Runte (2010) repeatedly notes that it was not the modern ideas of biodiversity conservation or ecosystem protection that gave rise to national parks. The initiators in the nineteenth century were more concerned with the picturesque impression of the landscape as well as the preservation of **natural wonders**. Similar to the European conservationists, the American ones seemed to be threatened by a total loss: in this case, however, not of an agricultural landscape that had grown over centuries, but of wilderness. The real-world equivalent of both landscape stereotypes seemed abundantly available barely a century before the respective conservation movements began around the mid-nineteenth century. The opponent in both cases was modern, technically and scientifically optimised agricultural and forestry use.

Like rock'n'roll and Coca-Cola, the idea of national parks began its triumphal march around the world; perhaps not because this idea was particularly good, but rather because it was American, i.e. it originated from the Western world power and therefore moved and fascinated many people, much like fast food, Disneyland and Hollywood. Moreover, the worldwide expansion of US-American nature conservation concepts also changed the direction of nature conservation in Germany and Austria, among other places: in addition to the protection of more traditional agricultural landscapes, the demand for wilderness in so-called process conservation now also occurred in Europe from the 1970s onwards. Where wilderness had already been completely destroyed in earlier generations, at least its restoration should be tackled: with the known longevity of trees, at least a work of centuries.

The USA made use of marginal landscapes which had not yet become settlement areas or which simply did not seem to be suitable in terms of natural space. Nature conservation thus occupied extreme landscapes to an even greater extent than in Europe (cf. Chap. 5). Since the Indian population was not believed to be capable of permanently transforming nature, it was defined as untouched. This demanded status of **untouchedness** was also incorporated into international objectives via the national park definition, such as those of the IUCN/WWF.

While the myth of unspoiltness was cultivated for the benefit of the marginal extreme landscapes, virtually all other original natural landscapes were converted into **extreme agricultural areas** in which small structures such as groups of trees or individual trees have always been almost completely absent. The fine composition of the landscape as an intensive mixture and interpenetration of cultural and natural elements, which is typical for European (and other Old World) cultural landscapes that have grown over centuries, is almost completely absent in the New World. Instead, we find monotonous slashes of one or a few crops reaching to the horizon. Whereas nature conservation in Europe primarily seeks to protect evolved and traditionally agricultural landscapes and biotopes that lie at the core of the old settlement landscape, in North America this protection of nature applies to the marginal and extreme landscapes, often desert and mountain.

This North American ideal of untouched nature is to be critically questioned. For example, the prairies with their huge herds of bison were (and in some cases still are) interpreted as nature; and this despite the fact that there is countless evidence that the Indian population cared for the bison on which they depended almost like pets and also created grazing grounds for them by pushing back the forest with the help of fire. Untouched nature seems to be less a reality than a projection in the US national parks.

On the other hand, in the U.S. national parks, tourist uses were and are not only permitted, at least in the peripheral areas, but are even expressly desired (Fig. 30.1). However, the US national park concept, like the internationalized one, fits in well with the demand for wilderness and with the idea of process conservation within nature conservation.

Fig. 30.1 Tourist uses have always played a major role in the concept of US national parks, alongside conservation concerns. In the Everglades, for example, free-living dolphins are trained for contact with tourists. (Source: Own recording)

However, such an orientation towards untouched nature becomes terrible when the national park ideal is uncritically transferred to Europe, where there is no longer any untouched nature and nature conservation turns primarily to the biodiversity of outdated agricultural landscapes.

It should be added that most states in the USA have created some **state parks**, most of which are smaller in area and contain less spectacular natural wonders, to some extent reflecting the idea of national parks at a regional level.

Brazil as an Example of an Emerging Latin American Country

There are several reasons why it is worthwhile to look at **Brazil** in terms of nature conservation.

On the one hand, Brazil includes more than half of the **wet forests of the Amazon**, which enjoy special attention in the context of international nature conservation as the **largest rainforest area** on earth. Less well known is the **coastal rainforest** (Mata atlantica), which is no less species-rich and even richer in local endemics. Originally it covered large areas of the east coast exposed to the trade winds, to the south as far as Rio de Janeiro, but today it has been destroyed or severely degenerated down to a few percent of its original area. Thus Brazil, like tropical America as a whole, possesses a biotic richness perhaps unmatched by any other major region in the world (Figs. 30.2 and 30.3).

When the first Europeans arrived on the east coast of Brazil in the sixteenth century, they found a "sea of forest". In order to build the first settlements, they literally had to clear into the forest, so this was first and foremost a process

Fig. 30.2 The amphibious habitat on the Amazon River and its tributaries supports several species of aquatic turtles, among others. (Photo: © Wenzel Halla)

Fig. 30.3 An Amazonian insect species of several hundred thousand: a cicada (Cicadidae) with protuberances on the abdomen, which presumably serve to emit sexual scent attractants (pheromones). (Photo: © Wenzel Halla/Stefan Hupke)

of **forest destruction** or forest displacement. It is therefore not surprising that the forest, which was abundant, was not considered to be of any particular value. Even the harvesting of wood, which the land needed for building houses and for commercial purposes, occurred incidentally in the context of these clearing processes and was not the result of preceding forest protection or targeted forestry as in the generally more densely populated Old World.

Well into the second half of the twentieth century, this indifferent, if not hostile, attitude towards the forest was reflected from the early colonial period. Significantly, the early critics of forest destruction came almost exclusively directly from Europe, as a contemporary print after C. Martius (Fig. 30.4) shows.

With the rise of **rainforest protection** as a powerful mainstream of the zeitgeist from the 1980s onwards, the Amazon rainforest also increasingly came into the focus of Western politics and media. For many years, the attitude of Brazilian politics and civil society contrasted with this international attitude, because the said colonial-influenced patterns of reception towards the forest were perpetuated. It

Fig. 30.4 Brazilian coastal mountains in the first half of the nineteenth century. The figure illustrates the slash-and-burn clearing of the rainforest for the then expanding sugarcane cultivation (in the picture on the right). (Bayerische Akademie der Wissenschaften 1995, p. 26, after C. Martius)

was not until shortly before the turn of the millennium that there was a gradual, and then increasingly rapid, adaptation to the international discourse. As a result, there is no longer any fundamental contrast between the foreign and domestic reception of rainforest protection. The contrast arises rather within the country between an urban-intellectual basic attitude in the cities of the southeast, which have found a connection to the Western discourse, and the persistent basic attitudes of the regional population of Amazonia, which, in view of the still prevailing availability of huge forest reserves, is not really able to see the need for rainforest protection.

The designation of protected areas, especially national parks, had to and has to be enforced against the resistance of the local population, even more so than in other cultures and regions. Demarcations are often ignored by the locals, which is easy to recognize in the satellite image, but often difficult to prevent due to the lack of accessibility by suitable transport routes on site. Some parks are so little respected that they can be considered "paper parks": protected areas that exist only on paper.

Until around 1970, **clearing in the Amazon** was only carried out on a relatively small scale. Around 95% of the lowlands were still in a near-natural state. The massive clearing that began thereafter has caused the forest to virtually disappear in some parts of the Amazon basin (Rondonia, Rio Branco) and has now multiplied the proportion of clearing in the whole of Amazonia to around 30%. If the present clearing trend is maintained, the Amazon lowlands as a whole will have become a largely forest-free terrain in just a few decades.

From the beginning, the clearances were based less on economic incentives or even necessities than on a state program to open up the interior of the country and thereby rise to become a major economic and political power. The unacknowledged model for this state building process was the

United States of America. In addition, there was the national idea of preempting the impending internationalization of the region by **colonizing Amazonia.** This idea could have been dismissed as completely misguided if such an internationally influential politician as the then French President Mitterrand had not made such an (unofficial) proposal as recently as the 1980s.

Tax breaks and direct subsidies, as well as the creation of infrastructure, were thus initially the driving force behind the settlement and economic development of Amazonia. In the meantime, these incentives have been successively reduced. However, the global demand for food and animal feed, and in particular for renewable energy raw materials, has increased to such an extent that, at least in the southern fringes of Amazonia, the production of soya, maize and sugar cane, but also of beef (Fig. 30.5), appears extremely

Fig. 30.5 Clearing island in Amazonia with extensive grazing (Indian Cebu cattle). Regrown secondary forest can be seen in the background. (Photo: © Wenzel Halla/Stefan Hupke)

profitable for the world market. The dynamic triggered by state support is thus tending to be replaced by the market-based incentive.

Overview

Soya is by far the most important animal feed in the world today. The EU is the second largest importer of soya, with 15 million tonnes annually. Eighty-eight percent of net soya imports into the European Union come from South America. Meat producers in Germany are among the largest purchasers within the EU, with around 4.5 million tonnes/year. According to WWF calculations, Germany needs an area of soy cultivation the size of Saxony-Anhalt for its soy imports in South America.

The soy boom is happening at the expense of biodiverse rainforests and savannahs. Between 2000 and 2010 alone, 24 million hectares of land in South America were converted, 20 million hectares of which were used solely for the expansion of soy cultivation. This corresponds to five times the size of the Netherlands.

(WWF Magazine 3/2017, p. 19).

At least in formal terms, however, Brazil cannot be reproached: since the 1970s, an abundance of nature reserves has been created in Amazonia, the largest of which are more extensive than any German state.

The Lula government had already enforced a legal regulation a few years ago, which was later relaxed (opponents would say: softened): each landowner is only allowed to clear a certain percentage of his land, which means that he has to carry out nature conservation on a certain part of his land himself, so to speak. Since then, new clearing areas have resembled a patchwork of used and natural areas. The disadvantage of this is that it is difficult for species to migrate between the isolated forest remnants. The advantage lies in a varied cultural landscape, which in many respects is reminiscent of old European agricultural landscapes. This is almost a novelty for South America, where otherwise, similar to the situation in North America, "primeval forests" reaching to the horizon on the one hand, and almost endless pure agricultural landscapes after clearing on the other hand are common.

400 New Species

In the Brazilian Amazon rainforest, a new animal or plant species is discovered every 2 days on average. According to the conservation organization WWF, 381 new species were found in 2014 and 2015. These include a species of jumping monkey (*Plecturocebus miltoni*), a stingray with a honeycomb pattern (*Potamotrygon limai*) and a pink river dolphin (*Inia araguaiaensis*).
(StZ v. 4.9.2017, n. dpa).

African Countries

Compared with South America, but also with North America, **tropical Africa** was colonially developed only late. The first east-west crossings (Livingstone, Stanley) did not take place until the middle of the nineteenth century, more than 300 years after the crossing of the South American continent of roughly comparable width by Orellana.

What the explorers and conquerors got to see in the interior of the African continent was a rich **world of large animals** that had no equivalent in the interior of South America, not even in the savannahs of the Brazilian campos, which were almost similar in area. In addition to the enormous numbers of individuals, such as the wildebeest herds, the number of species is particularly astonishing. Thus there are (according to today's knowledge) in Africa two elephant species (Elephantidae), two rhinoceros species (Rhinocerotidae), two representatives of the giraffe

(Giraffidae) and two hippopotamuses (Hippopotamidae), for which there is no equivalent at all in America. The equidae, which were also absent from America before the arrival of the Europeans, are present in Africa in the form of several species of zebras and wild asses. Among the large carnivores there are several species of large cats lion (*Panthera leo*), leopard (*Panthera pardus*), cheetah (*Cinonyx jubatus*), as well as of the dog-like (Canoidea) and hyenas (Hyaenidae). Represented in an almost unmanageable variety of species, however, are the antelopes, an animal-systematically very heterogeneous group of ruminants that sometimes resemble horses externally (horsebucks), sometimes seem similar to cattle like the wildebeest (*Connochaetes* spec.) and other cow-antelopes, sometimes resemble deer or goats like the dainty gazelles.

To the arriving Europeans, this diversity of African wildlife initially appeared as a pure product of nature (e.g. Adams and McShane 1996, p. XIII) and was thus not dissimilar to the reception of nature in North America. If anything, this natural wealth had to be protected from the indigenous blacks (Shetler 2007, p. 181 ff.), who, after initial declines as a result of the mostly rigid colonial system, soon began to multiply considerably as a result of modern scientific medicine, which was also induced by the Europeans.

The best means of protecting wildlife rich in species and individuals seemed to be **large-scale protected areas** in the form of **national parks**, which were soon designated as priority wildlife sanctuaries by the colonial powers. In the process, traditional settlement and hunting rights came to a halt; local people knew use rights, but traditionally not land ownership in the European sense, and were effectively dispossessed (Adams and McShane 1996, p. XV). Even more than in North America, national parks were enforced against the indigenous population (the corresponding North American population had already been relocated to reservations before, exterminated, or died out on their own due to diseases introduced by Europeans).

When the rural African population concerned resisted and insisted on their ancestral traditions of use, they were quickly criminalized as poachers (cf. Grzimek and Grzimek 1961; Adams and McShane 1996, p. XV).

It is interesting to note that even large-scale protected areas (the largest of which correspond to average German federal states) rarely function when left to their own devices. This is particularly evident in South Africa's Kruger National Park, where after a few decades the multiplying elephant herds began to destroy the woody plants and trees of the savannah in excess. There had obviously become too many elephants under the protective measures.

But too many elephants—how can that be in an ecosystem that should actually regulate itself? After all, every species has its natural enemies that it keeps short. Really every species? In any case, the huge and defensible elephants are rarely attacked by lions and certainly not by other predators of the African savannah, not even the young ones, because they are effectively protected by the old ones. But then who was able to prevent the harmful mass reproduction of elephants in the past? A sensible answer is: the Africans themselves, by hunting the elephants. The elephants (Elephantidae) as a species responded in their own way to this hunting pressure, which may have lasted for hundreds thousand of years, by increasing their reproduction. The natural balance of the savannah, seemingly untouched by man, cannot do without man as a control. The fact that excessive hunting pressure, especially from modern firearms, can in turn endanger large animal species should not be doubted.

The enormous **species richness of** the African large animal world is virtually proof that early human settlement need by no means have a negative effect on the animal world. In all other continents or large islands such as Madagascar or New Zealand, human existence lasted much shorter and came literally overnight as a result of rapid settlement. Thus it came about that everywhere outside Africa the Pleistocene world of large animals was largely wiped out about 6000–10,000 years ago (Pleistocene overkill; Martin and Wright 1967). In Europe, this affected giant deer and mammoth, forest elephant, forest rhinoceros, steppe rhinoceros, cave lion, cave bear, cave leopard and cave hyena (and many other species).

Because of the slow growth and development of young animals, especially in elephants, the generations follow one another at long intervals, as in humans, thousands of years are required before a true new species is formed. However, human influence in Africa can be traced back that long, with fire (cf. Wrangham 2009) and hunting being the significant influences. While fire influences the vegetation and above all pushes back the forest, thus favouring the open savannah and grasslands as a forage base for the large herds of wild animals, but thus also for predators, hunting directly influences the game populations.

In any case, through the genetic process of co-evolution, animal life, plant life and man have evolved together in Africa. In a sense, African man has become the creator of the surrounding nature, its animal and plant species. This cannot be said for any other continent.

Contrary to this view, African nature is today widely seen in the Western media in great contrast to African people. In a certain sense, African nature has been de-Africanised, i.e. very close attention is paid to whether Africans deal with this nature appropriately and correctly. If not, this quickly becomes the subject of a session of the German Bundestag, which, by virtue of its territorial competence, should be more interested in Central European than in African matters. National and international bodies outside Africa are campaigning for a ban on the ivory trade and for the protection of African elephants. When a group of African states submitted a motion for the protection of the North Sea herring at a meeting of a UN body a few years ago, this was dismissed by the European side as “cheap polemics” (!). Why, in fact? Weren’t the stocks of herring in the North Sea also in alarming decline?

In principle, Europeans have always regarded African nature as their very own possession. Neither, from this perspective, should “the Chinese” secure the African trade in raw materials, nor should Africans themselves decide about their wildlife.

Although the colonial powers forbade the indigenous peoples the traditional use of game stocks within the protected areas as **poaching**, at least their elevated governors preferred to hunt African big game. Often the game reserves were established precisely for the **big game hunting of** the Europeans. After decolonisation in the 1960s, the privileges of the European upper classes were transferred to the new African elites, most of whom were socialised in urban areas and were equally uncomprehending of rural claims to use the land. Today, big game hunting, like all elitist forms of tourism, has lost both spatial effectiveness and economic significance and has been replaced by **safari tourism** in which photographs are the actual hunting trophies.

Different countries have taken different paths in this safari tourism. While Kenya relies on low-cost mass tourism combined with beach holidays, Botswana, Namibia and Zimbabwe have developed upmarket tourism that brings

Greed for Ivory Decimates Animal Population Extremely

Elephant conservationists are sounding the alarm: the number of African pachyderms is much lower than previously thought—and is declining by 30,000 animals a year due to increasing poaching. This was the conclusion of an unprecedentedly thorough elephant census conducted over the past two and a half years in 18 wildlife-rich states in Africa. The census revealed that only 350,000 savannah elephants remain in the 18 countries surveyed: Experts had previously assumed half a million.

[...]

The only bright spots have been in the south of the continent—in Botswana, Zimbabwe and South Africa, where 60% of all African savannah elephants now live. With more than 130,000 individuals, Botswana has the continent's largest pachyderm population, presenting the arid country with new ecological challenges. Satellite tracking has revealed that elephants threatened by widespread poaching in Angola are seeking refuge in relatively well-protected Botswana: Now the population density of jumbos there is so high that tree cover is suffering almost irreparable damage.

Before colonisation by Europeans, around two million elephants are said to have lived in Africa. Their numbers declined steadily until conservation efforts were able to stop the disastrous trend from the 1970s onwards. But poaching has been on the rise again for the past decade, attributed to the growing affluent middle classes in China and Vietnam. There, ivory is used as a raw material for jewellery and is considered a remedy.

The illegal trade in tusks is in the hands of highly professional gangs who use local helpers to kill the animals. The elephants are killed with high-calibre hunting rifles, but often also with Russian rapid-fire rifles or poisoned arrows, causing enormous pain. In the former civil war country of Angola, it is even said that the animals are blown up with land mines or shot at with grenades.

(Johannes Dietrich in StZ v. 2.9.2016).

From World Heritage to Construction Site?

Africa's largest protected area, the Selous Wildlife Reserve in Tanzania, is threatened with transformation into a huge construction site. According to an agreement signed between Tanzania and Egypt in December 2018, an Egyptian consortium is to start building a dam there as early as June 2019 to produce electricity. With it, the heart of the world's natural heritage with an area of about 1200 km^2 would disappear in a reservoir. In view of the acute threat situation, the German Bundestag debated the issue in plenary on 17 January 2019. In a joint motion by the CDU/CSU and SPD, which was also approved by the Greens, the MPs call on the Federal Government to work to preserve the unique protected area and to find sustainable alternatives to electricity supply.

WWF Magazine 2/2019.

Author's note: The article particularly shows the national centricity of the political actors in Germany (as far as the thought patterns are concerned) on the one hand, and on the other hand also the claim to world competence. One may wonder whether they would have reacted in the same way if the dam project had affected a Western industrialized country. Only shortly before, Trump's government very much relaxed nature and environmental protection laws with regard to oil production in Alaska, without the German Bundestag having debated this in a session. Factually-territorially, the German parliament has no jurisdiction over Alaska or Eastern Africa. Even the fact that the German government has financially supported this reservation in the past seems insufficient as a justification for this formal interference. Apparently "donated" funds for "development cooperation" thus become a vehicle for external interference for the societies concerned and are thus dearly bought.

On the other hand, one has to ask why renewable energy from hydropower, widely preferred as "sustainable" by political bodies in Germany, should not be sustainable in Africa.

Incidentally, a reservoir in the centre of the huge protected area could well stabilise the animal population, because in the tropical, alternating humid climate the availability of water at the end of the dry season is a minimum factor, especially for the preferred large mammals.

fewer travellers into the country at higher prices and often leads to a similar regional value added with higher average expenditure per guest. In both cases, the respective economies certainly benefit from the increase in tourism in the national parks. Regionally, on the other hand, the effects remain comparatively small in many cases, as often only the simple work in the hotels remains for the local population. Almost nowhere has the rural population, deprived of its traditional uses, been truly reconciled with the nature reserves they have been deprived of.

It is therefore not surprising that the pressure from poaching continues. The focus is often not only on supplying local markets with bushmeat, but increasingly on supplying Asian markets with rhino horn, which is believed to have great healing properties in traditional medicine. Because a lot of money is paid for the rhino horns, outside of southern Africa the two African rhino species have been almost universally destroyed or brought to the brink of extinction. Recently, however, poaching of rhinos has increased enormously in South Africa as well.

Nevertheless, it can be stated that in no other continent are such large areas designated as protected areas as in the savannah landscapes of eastern and southern Africa, partly due to the obvious advantage in terms of foreign exchange. In some countries, the proportion of protected areas exceeds 20% of the total national area.

On the other hand, outside these mostly generously designated areas, strong population growth and increasing agricultural development are very rapidly consuming the African "wilderness" that covered almost the entire country about a hundred years ago (Fig. 30.6); the near-natural landscape of the protected zones is becoming isolated. But this phenomenon is not specifically African. It occurs only with a historical delay and affects a particularly rich large animal world.

Fig. 30.6 Small-scale settlement has consumed the forest areas of the Ethiopian highlands, which were extensive only a few decades ago, except for a few remnants that were left standing here for reasons of firewood supply and as shade trees for the cattle. A varied cultural landscape has developed with small arable areas between extensive cattle pastures, which was also prevalent in large parts of Central Europe until the nineteenth century (cf. Chap. 2, Fig. 2.1). However, these areas are unsuitable for the species-rich African big game. (Photo: © Kornelia and Friedrich Gervé)

India

In some respects, the historical situation in **India** is comparable to that in Africa. Both regions were dominated by European colonial powers until after the Second World War. However, both also have their own distinct pre-colonial traditions. In addition, both spaces are on the whole not dominated by a monotheistic religion, but have allowed something like religious "wild growth". Until the present animistic traditions could live on, which in the case of India are unbrokenly connected with "book religions", but which also draw from these folk-religious traditions. All these influences have helped shape the understanding of nature.

If one looks at a historical account of the history of nature conservation, of which there are several at least as introductory chapters in monographs devoted to nature conservation, the beginnings of nature conservation are almost always anchored exclusively in Europe and North America. Like almost all modern ideas, following these accounts, the idea of conservation spread worldwide from these centers: Europe (with its North American offshoot) as the teacher, the rest of the world as the learner.

This is in contrast to the fact that there have been, and still are, protected areas in both Africa and India since time immemorial. Only in these cases, one does not speak of nature conservation. They are mostly old forests that are exempted from use. In the cultures concerned, this **exclusion from use** is not justified in terms of nature conservation or even ecology, but in religious terms. They are *sacrified forests*, dedicated to a certain deity or spirit of nature. In fact, however, these areas usually fulfil all the requirements of a nature reserve. It is forbidden to fell trees in these forests. However, the removal of *minor* forest *products* such as herbs and berries is often permitted. This circumstance seems decidedly modern, because the protection of forests through the use of by-products has been widely discussed in the past decade (for a summary, see Shackleton and Shackleton 2011).

In addition, Western (i.e. international) influences have of course also arrived in India, for example in the form of the designation of national parks. In many cases, these emerged from British **game reserves**, which is comparable to the situation in Africa. The formally educated strata of Indian society are for the most part very open to this conservation concern. This is also due to the fact that **Hinduism** does not know the strict separation into animal and human as it characterizes the biblical idea of creation in this form. Animal and human are rather staged manifestations of the same soul and can, in the sense of the transmigration of souls, pass over into the other form of life after the death of the one. In this world view man is not necessarily born to rule and also does not possess a divine order to "subdue the earth".

The fact that the animal world is regarded for religious reasons rather as a fellow creature, without a quite principled separation between man and animal, suggests also in the Indian cultural context the already from Europe known mixing up of animal protection as protection of the rights of the animal individual with nature protection as protection of species and habitats.

Today, India is covered by a network of **national parks** and other protected areas in which, as in many African countries, a remarkable amount is being done to protect nature. At the beginning of the millennium, the proportion of protected land was just under 5% of the total national area (NEGI 2008, p. 236), which is slightly higher than in Germany. Considering the high population density and the strong pressure of agricultural use in India, this proportion is in itself astonishing.

Compared to Africa, however, safari tourism is of much less economic importance, although it is also growing. In contrast to many African countries, local people make up the majority of tourists in national parks and other nature reserves in India. In India, too, the idea of nature conservation tends to be supported by the urban and educated classes, which are, however, the dominant group in society as a whole. Here, too, there are the conflicts between nature conservation and a land-hungry rural population that are already familiar from Africa.

Bordered to the north by the barrier belt of the Himalayas, the Indian subcontinent has a high number of **endemics**. Particularly in the mountainous areas of northern India, but also to some extent in the southern Indian Western Ghats (Figs. 30.7 and 30.8), near-natural habitats have

Fig. 30.7 Semi-evergreen forest in the southern Indian Western Ghats. Unlike in the actual tropical rainforest, some of the large trees lose their foliage during the winter dry period. (Source: Own recording)

been able to survive. Since the predominant religion of Hinduism actually precludes hunting, remarkably large populations of large forest animals have survived to the present. This is true for large populations of the Indian elephant, whose population is still estimated at tens of thousands for all of India. But also half of the world-wide tiger existence is allotted to India, altogether probably 2000–3000 animals. This is astonishing, since in Germany the reintroduction of much smaller predators meets with great resistance—even though India is more densely populated and more dependent on land use than Germany.

In addition, as in other (alternating) humid tropical regions, an immense diversity of birds, reptiles (two representatives are shown in Figs. 30.9 and 30.10), amphibians, freshwater fish and arthropods is to be protected in India. Figure 30.11 gives an example of a high species density of mammals depending on the location.

Fig. 30.8 Nilgiri langur (*Semnopithecus johnii*). Although the animals can weigh more than 10 kg, they are almost pure tree dwellers. Since they need strong branches (see picture), they depend on natural old forests. In the southern Western Ghats, these cover only a few percent of the original area and are also very fragmented spatially. Although the population is still quite considerable (the species can be observed regularly, as here during a geography high school excursion), the local endemic, which occurs only here, is considered potentially endangered. (Photo: © Annelie Bayer)

Fig. 30.9 The Indian Star Tortoise (*Geochelone elegans*), with its intense shell patterning, is one of the world's most beautiful tortoise species and is therefore, although still common in many regions of India, potentially threatened by collection. (Source: Own recording)

Fig. 30.10 The marsh crocodile (*Crocodylus palustris*) has disappeared almost everywhere in India as a result of deliberate persecution and road traffic, surviving only in larger protected areas. (Source: Own recording)

Fig. 30.11 If the visibility of the terrain is poor and the animals are predominantly active at night, it is quite difficult to detect individual species. Indirect observations such as footprints can help. Here, at the edge of a reservoir in a South Indian national park, several predators followed the lakeshore one after the other during the night in search of prey animals coming to the water to drink: a sloth bear (*Melursus ursinus*; tracks in the background) as well as presumably a jungle cat (*Felis chaus*; small and flatter impressions, rather in the rear foreground). In between, impressions pointing to a dog-like predator, and since there are at least two animals and the depth of the indentation suggests a heavier weight, probably to the red dog (*Cuon alpinus*). (Source: Own recording)

Catastrophic Vulture Deaths in India

In India, Pakistan and Nepal, an animal tragedy of enormous proportions is taking place. Within only about 10 years, the former army of millions of Indian vultures, Bengal vultures and slender-billed vultures has melted down to small remnants. Depending on the species, only 1–3% of the birds have survived and there is still no end in sight.

[...]

The affected birds show gout-like symptoms and eventually die of kidney failure. While initially a still unknown viral disease was suspected, researchers have since discovered that the drug Diclofenac is the main cause of the mass death. This anti-inflammatory drug, which originated in human medicine, has also been used in veterinary medicine since the 1990s in India, Pakistan and Nepal—especially in cattle. The vultures ingest the active ingredient via the domestic animal carcasses on which they feed (www.nabu.de/tiereundpflanzen/voegel/international/03.530.html, news item dated 29.3.2005, accessed 17.9.2018).

31

Nature Conservation in the Third World: A Pillar of Neocolonialism?

In May 1990, the Enquete Commission "Precautionary Measures for the Protection of the Earth's Atmosphere" of the German Bundestag presented a work of about 1000 pages, dedicated to *the protection of tropical forests—an international priority task*. Initiatives supported by governments and parliaments in Europe and North America took up the protection of the African elephant. The protection of the African savannah and its wildlife have always been of particular concern to Europeans (Chap. 30); Grzimek B and Grzimek M *Serengeti darf nicht sterben* (1959), for example, was translated into dozens of languages and marked the beginning of the designation of large-scale nature reserves in Africa. Europe's media, Europe's citizens, Europe's governments and parliaments see themselves as responsible for the protection of nature in other parts of the world and climatic zones as well.

In the mid-1980s, the then German Chancellor Helmut Kohl travelled to Brazil, had himself flown to one of the newly created rainforest conservation areas in Amazonia,

K.-D. Hupke, *Nature Conservation*,
https://doi.org/10.1007/978-3-662-66159-8_31

and there made a commitment to the protection of the jaguar (*Panthera onca*). At first this cost nothing, no tax money at any rate, because as is well known the German Chancellor is not territorially responsible for Amazonia. However, one must concede to Chancellor Kohl that in the years that followed he set up a joint German-Brazilian research and action programme for the Amazon rainforest and made several million DM available for this purpose.

What is striking, however, is the fact that at the same time Helmut Kohl could have done much more to save an endangered cat species, and much closer to home. An initiative was launched to reintroduce the lynx to the Black Forest, perhaps the largest closed forest landscape in Germany. However, an alliance of farmers and hunters prevented this reintroduction at the time. Concerns were expressed in particular that the lynx could lead to losses of deer and lambs.

The fact that Chancellor Kohl massively demanded (and finally even financially supported) the protection of the much larger predatory cat in the Amazon, but at the same time was unable to enforce the protection of a comparable species in his own country, throws a spotlight on the assertiveness of nature conservation interests in general.

It was no different with Helmut Kohl's friend on the other side of the Rhine. On the occasion of a trip to Brazil, the then French President Mitterrand called for the **internationalisation of Amazonia**. This immediately sparked a storm of protest from politicians and the media in Brazil. The age-old Brazilian view, always characterized as irrational by European observers, that only a populated Amazon would secure Brazil this part of the country as a permanent possession, was given new impetus—and above all a sense of reality. Imagine, conversely, a foreign state guest in Paris demanding the internationalization of the Louvre—or any other French institution or part of the country! The reversal of perspective, as attempted by the Africans in the protection of the North Sea herring (cf. Chap. 30), impressively demonstrates how much we expect Africans and South Americans to do in terms of **territorial interference**—especially in nature conservation.

Often our language is already treacherous. Almost every child knows that the forest areas in almost all tropical countries are drastically decreasing. But what is a forest? If the term "**forest**" also includes forests that are far removed from nature and consist of planted stands (which are counted as forests in almost all our statistics), then why aren't the planted plantations of oil palms or rubber trees in Malaysia and Indonesia also forests? As a matter of course, we use a different concept of forest for the tropics, one that is oriented towards the presumed original state. If we were to apply this strict concept to Germany as well, Germany would have no (or almost no) forest (left).

In general, Europeans (and the North Americans who are culturally and genealogically derived from them) like to see untouched nature in the countries of the so-called **Third World**. While in Germany the use of the forest, for example in the forester's profession, is often almost idealized in a unity of forest user and forest protector (shouldn't it be conceivable that there are also conflicting interests?), any use of "untouched" nature in tropical forests is simply rejected. This has obscured the fact that even the rainforests of the Amazon show many signs of a pre-European human imprint, which research in recent decades has only gradually begun to uncover (e.g. the existence of anthropomorphic soils (Lehmann et al. 2004) or the remarkable proportions of fruit-bearing trees in local forest images). Not entirely without reason, some researchers speak of the ancient cultural landscape of Amazonia when referring to the Amazonian rainforest.

But there are other aspects to the relationship between the richer and poorer parts of the world. At least for forest protection, relatively much is being done in the wealthy economically developed societies at the moment. In recent decades, for example, the naturally dominant slow-growing deciduous tree species have been somewhat favoured again. This is at the expense of wood production, which, among other things, covers the still growing demand for pulp and paper. For these reasons, Germany is the largest European importer of paper and its precursor pulp. This means that if domestic forests are to be protected, exotic forests must be used more to meet domestic demand. For this reason, in addition to Nordic countries such as Finland and Canada, more and more wood is also coming from South America, where, for example, the previously open savannah areas of the Brazilian campos, which covered almost half of this giant country just a few decades ago, are increasingly being taken over by eucalyptus plantations for wood production, with the destruction of near-natural habitats. A comparable development can also be observed in the neighbouring countries of Argentina and Paraguay. As in the case of renewable energies (Chap. 33), the same can be said here: Nature conservation (and environmental protection) in the rich industrial societies has its price—the poorer countries have to pay it.

Neocolonial relations are characterized by **inequality**. This is no longer predominant today, or at least not exclusively an inequality of the direct means of power. It is, however, reflected in an inequality of applied standards and norms. Africans, Asians and Latin Americans are mostly aware of this inequality; they have grown up with it. We, on the other hand, are not; at least not on the losing side.

32

Nature Disappears, Nature Conservation Comes?: On the Alibi Function of Nature Conservation and Nature-Protected Areas

Nature that is not fundamentally threatened by humans does not need nature conservation. In this respect, it is not surprising that, for example, in the vast Amazon, nature conservation was not an issue until the 1970s; at least not for the planning authorities. The seemingly untouched forest areas were still huge, clearing limited to about 5% of the rainforest area. What sense, then, would it have made to designate a more or less arbitrary part of the vast sea of forest as a nature reserve or a national park? Outside the boundary demarcation, which could hardly be determined in the terrain anyway, there would have been a comparably diverse and untouched nature.

This changed when, from about the mid-1970s onwards, the end of the huge rainforest areas seemed foreseeable for the first time. It was not the case that, at the rate of deforestation then (and now), the rainforest would not have lasted for a few more decades. But the end of the huge rainforest areas in the temporal context of about one human lifetime

K.-D. Hupke, *Nature Conservation*,
https://doi.org/10.1007/978-3-662-66159-8_32

was within reach. The prerequisites for this were technical innovations such as chainsaws and bulldozers, and, from around the beginning of the new millennium, a rapidly growing global demand for agricultural products as a result of rising living standards in many countries and the growing demand for renewable energy sources.

The designation of large national parks or other protected areas is almost always linked to deforestation plans in the spatial surroundings of the newly created protected areas, which always exceed them in size by several times. The designation of **large nature reserves**, always hailed as a success by the active NGOs, becomes an accompanying factor in the rampant **destruction of the rainforest**, namely on all non-protected areas in the wider surroundings. Comparable linkages between protection and increased deforestation, both in historical and regional interactions, are also known from the Southeast Asian region (Malaysia, Indonesia). Nature conservation in this context is not only a success story, but also an indicator and accompanying factor for nature destruction. Likewise, the token character of nature reserves in order to be able to destroy nature outside and on a much larger scale is often more than obvious.

Within the nature conservation legislation of **the** EU, this **linking of nature destruction with nature conservation** is even expressly cultivated. For example, the new wording of the German Federal Nature Conservation Act of 2009 explicitly enforces the designation of **compensation areas in the** case of nature-destructive measures, for example in connection with the designation of new building areas (Fig. 32.1). An area whose nature conservation value is impaired by a development measure must be upgraded by another area in the vicinity. In comparison to the above-mentioned examples from Amazonia or Southeast Asia, however, German nature conservation law requires that the compensatory areas be comparable to the destroyed

Fig. 32.1 "Ecological compensation measure" in favour of upgrading a replacement biotope; example Heidelberg Bahnstadt. Buildings for start-ups and apartments were constructed here on a former freight station and marshalling yard site. A marginal former railway line with an approximately 2 m high embankment was upgraded with dry stone walls, rock packs and dead wood. As part of the biotope maintenance elsewhere anyway accumulating mown material, which in autumn automatically contains plant seeds, was spread here and thus stimulated an initial vegetation of local species. After the dry stone walls had been erected, the wall lizards, which had spread over the area to be built over later, were relocated to their new biotope, which was also very well received. In addition to the nature conservation purpose, the apartments being built were also enhanced by the proximity of a double green strip: rather close to the houses (top left in the picture) by more horticulturally designed plants, further to the right and in the foreground by the aforementioned compensatory biotope; in between is a connecting cycle and footpath. (Source: Own recording)

areas—both in terms of the quality of the areas concerned and in terms of their size. This means that the connection between the destruction of nature and the enhancement of nature in German nature conservation law loses its token character, and genuine compensation can be achieved. The causal link between nature destruction and nature conservation/nature enhancement nevertheless remains very close and leaves the nature conservationist feeling a certain unease.

Another question is how to compensate destruction of nature in principle with nature conservation measures.

Let us put ourselves in the spatial situation of, say, Western Amazonia. Rainforest, which is considered primary, is continuously being cleared there. To compensate for this, not quite as large, but equally extensive nature reserves have been designated. An effective compensation for the destruction of other areas?

Not at all. A protected status has a provisional character, nothing more. It is decreed by government decree or by law, and it applies as long as that law exists. But the legal basis of a protected area can be changed at any time. This is by no means hypothetical. In Brazil, the nature conservation status of some areas has been revoked in recent years, and in Ecuador and Peru, at least, forms of use contrary to the status, such as the extraction of oil, have been approved in nature reserves.

The clearing of a rainforest area, on the other hand, is of a completely different factual quality. Regardless of how our attitude to rainforest and conservation issues will develop in the future: A primary rainforest cannot be restored, at least in human time scales. Thus, the destruction of the rainforest cannot be stopped in principle by nature reserves, but only delayed (depending on future changes in the political and economic perspectives of the relevant societies). This is not nothing. But it is not "protection for all time" or even a fundamental compensation for the destruction of nature. If you set nature-protected land against nature-destroyed land, the imputed basis of the calculation is wrong.

To a certain extent, this also applies to "native" Central European nature conservation. The compensation of extensive development measures by nature conservation enhancement of other areas is often celebrated as great progress. What is overlooked is that overbuilding measures can be assumed to have a certain historical finality. Once an area has been built over, it does not quickly revert to a near-natural habitat.

What about the compensation measure? For example, a previously agricultural area is bought up and maintained as an extensive meadow. The contractual commitment foresees a term of perhaps 40 years, which is to be pre-financed by the investors of the consumed area. That is certainly a long period. But after that? In 40 years, the pond created as a compensation measure will have silted up and the extensive meadow will have to be maintained. There is no money for that now. As in the case of the nature reserve in Amazonia, the following also tends to apply here: the compensatory measure is only temporary, the destruction of nature, on the other hand, is definitive in human time scales.

Brazil Releases Huge Nature Park for Mining

The Brazilian government will allow mining in a previously protected region in the future. This is the result of a decree issued by former president Michel Temer.

The National Park Reserva Nacional do Cobre e Associados (Renca) in the northern states of Amapa and Para covers around 46,000 km^2 and is thus larger than Denmark. Large deposits of gold and other minerals are suspected there.

"The aim of the measures is to attract new investment, create wealth for the country and jobs and income for society," the Reuters news agency quotes a statement from the responsible Ministry of Mines and Energy. All this is happening "on the basis of sustainability rules".

(Spiegel.de/wirtschaft/soziales/amazonas, ges. 25.8.2017).

33

Nature Conservation in Times of Climate Change

Reichholf (2009, p. 114) can be agreed with in principle when he states:

> In any case, however, climate warming in the 150 years since the mid-nineteenth century changed far less in our forests than forestry. Where something like natural forest still grows, there are hardly any traces in the species composition that indicate climatic changes.

However, the threat does not lie so much in the extent of warming that has occurred so far, which should correspond to the fluctuation ranges that have occurred so far in the post-glacial period. In this respect, one must be cautious about wanting to "observe" the effects of climate change now, given the small effects that can be expected at the moment. Chapter 7 attempted to show that the increase in many "Mediterranean" orchids cited as evidence of climate change effects can be explained solely by a change in the

K.-D. Hupke, *Nature Conservation*,
https://doi.org/10.1007/978-3-662-66159-8_33

"management" of the corresponding open land types, without having to refer to climate change. The extinction of the bog saxifrage (*Saxifraga hirculus*; Fig. 33.1), which is considered an ice-age relict in Central Europe, is similarly open to interpretation in Germany as a possible consequence of climate change; it could also be a simple consequence of eutrophication caused by atmospherically introduced nitrogen compounds. The problem lies rather in the temperature increase to be expected in the future, the human origin of which, although not strictly provable, can be assumed as probable due to the close correlation with the emission of climate-changing trace gases.

If the global increase in carbon dioxide (but also methane, nitrous oxide, etc.) concentrations continues unchecked (or slows down only slightly), the global mean temperature is expected to rise by at least 3 °C, but more

Fig. 33.1 Bog *saxifrage* (*Saxifraga hirculus*): an inhabitant of northern German and pre-Alpine fens, considered an Ice Age relict. Extinct in Germany as a result of global warming?. (Wikipedia)

likely by 5 °C or more by the end of this century (https://science2017.globalchange.gov/). This would correspond to a thermal shift from southern Germany roughly to the Côte d'Azur. It is unlikely that this would be without consequences for species dynamics, with thermophilic species tending to increase and thermophobic ones to disappear. How quickly such a migration would succeed seems uncertain; especially since the rapidity of climatic change as well as the difficulty of migration for many species due to the infrastructurally predetermined islanding of the areas should generally make migrations more difficult. As these are highly complex and competitive ecological relationships, it is also almost impossible to draw up clear scenarios for the future. In general, an increase in overall species diversity due to climate change seems possible, at least for southern Germany, especially since warmer climatic zones generally have higher species numbers under comparable precipitation conditions. However, it also seems conceivable that the infrastructural impediments to migration and the slowness of many migration processes will not initially change the species composition much at all. However, a certain shift in species frequencies is to be expected due to the high temperature difference (Figs. 33.2, 33.3 and 33.4).

Possibly (even) stronger than the warming could be a possible shift in precipitation on the composition of vegetation and fauna. Some climatologists expect an increased frequency of hot and dry summer months in Central Europe, approximating the summer-dry climate of Southern Europe (which is already beginning in Southern France at the moment), in this case combined with an increase in precipitation in the winter months caused by increased cyclonic precipitation, but also probably in the transitional seasons. Both effects together make it seem possible that in less than a century we will be surrounded by an evergreen hardwood forest (in the manner of natural

Fig. 33.2 As a warmth-loving pioneer species, the red-berry bryony (*Bryonia dioica*) is found in southwestern Germany almost exclusively at low altitudes below 500 m NN. The extremes reach up to over 700 m (Sebald et al. 1993). For some years now, however, it has been found increasingly often at the upper limit of its previous distribution. (Own recording)

Mediterranean forests) as potential natural vegetation. The most important representative, the holm oak (*Quercus ilex*), already thrives quite well in the green spaces of "climate-favoured" German cities.

Fig. 33.3 Presumably as a consequence of decreasing frost extremes, the laurel cherry (*Prunus laurocerasus*), *which is* frequently planted in gardens, is increasingly also found in the forests of climatically mild regions of southwestern Germany. As an evergreen shrub or tree, the laurel cherry is already a representative of the climatic subtropics and native to Asia Minor. As the very cold February 2018 near Heidelberg showed, severe frost damage occurs below about -8 °C, especially to the young branch ends, but this could be compensated for by increased growth in the following vegetation period. The extreme frosts of up to about 20 °C below zero, which still occurred occasionally until the 1960s, would probably kill the shrub. (Own recording)

Fig. 33.4 The praying mantis (*Mantis religiosa*) is the only Central European representative of a primarily tropical insect order. In Baden-Württemberg for a long time only native to the extreme southwest (Kaiserstuhl-Freiburg area), it has been migrating northwards for several decades and has meanwhile reached the Karlsruhe/Heidelberg area via the Ortenau region, which is probably due to the warmer and longer summers (the species is relatively insensitive to winter cold). (© Wenzel Halla)

Under the circumstances of the presumed future climate change, more and more voices are calling for active adaptation to a development that may in any case no longer be preventable. This includes the "Arbeitsgemeinschaft Deutscher Waldeigentümer" (AGDW), the politically influential association of large private forest owners. According to their ideas, deciduous tree species that require a lot of water should increasingly be replaced by Douglas fir plantations (*Pseudotsuga menziesii*; Fig. 33.5), a coniferous species originally from North America that has proven to be very adaptable to all conceivable climatic extremes. On most forest sites in Central Europe, this tree species is accordingly the most vigorous, i.e. it generates the highest forestry yields. Since Douglas fir wood also proves to be exceptionally weather-resistant in outdoor applications, nothing should stand in the way of an ever-growing market segment for this species of wood. According to the AGDW, German forests should be "made fit for climate change" with the help of Douglas fir.

From a "nature conservation" point of view, an increasing proportion of coniferous wood appears to be questionable, after the Central European forests have been largely converted into coniferous forests since the nineteenth century and, in contrast to this, a slight tendency towards near-natural deciduous wood has been established again in recent decades. Conifers produce a long-lived litter on the forest floor which, in contrast to deciduous trees, decomposes only slowly, seals it and, moreover, produces an acidic forest floor with a decreasing pH value. In the times of the German "Waldsterben" (1980s), acidic forest soils were considered a major cause of tree disease and death in areas with low-calcareous geological substrates. In comparison,

Fig. 33.5 The Douglas fir, a vigorous tree species—even under the conditions of climate change?. (Own recording)

deciduous trees (somewhat restricted: the copper beech *Fagus sylvatica*) mostly form a rapidly decomposable humus (= gauze) that mixes well with the mineral soil and has an almost neutral pH value, which enables a soil life rich in species and individuals and which is suitable for species-rich plant communities of early bloomers (cf. Semmel 1977, p. 20 f.).

A corresponding criticism of the Douglas fir, which is clearly preferred for economic reasons (Czybulka et al. 2018), was countered by the AGDW ("Die Waldbesitzer") with the argument that in the pre-glacial period (i.e. until about 2 million years ago), the Douglas fir can also be traced as a species in Central Europe and only became extinct due to the ice age. In this context, the planting by humans could almost be seen as compensation for nature being "damaged" by natural temperature fluctuations. Once again, it becomes clear that "nature conservation" (understood in the literal sense of "protection of nature") can be understood to mean a great many things; sometimes also very contradictory things.

When it comes to climate change, however, it is important to look beyond Central Europe and consider the global consequences. Dry zones have tended to develop along the tropics of the sun's zenith in early summer, which only exceptionally, for example in oases with spring water outlets, allow for more intensive plant and animal life. These arid core zones are therefore by no means insignificant from a conservation point of view, but appear to many conservationists to be of less central interest in comparison with the very species-rich tropical rainforests and in comparison with the alternately humid tropical savannah zones with their very species-rich world of large animals, especially in Africa (Chap. 30). But it is precisely these species-rich savannah landscapes that, in the event of a further increase in temperature, seem to be particularly constricted by an ever-moist forest zone extending from the equator and a year-round dry zone likewise extending, in this case towards the equator. The actual temperature increase does not seem to play a major role in the tropics, unlike in the scenarios with extremely heated polar regions and corresponding effects such as glacier melt and thawing of permafrost.

If one takes a very long-term view of climate events, across the major geological epochs so to speak, it is noticeable that the entire Tertiary period (today: Paleogene and Neogene), for around 65 million years, was significantly warmer overall than the most recent period of around 2 million years. Nature has adapted to these temperature fluctuations in a recognisable way. So no reason to worry about warming?

Not quite. During past periods of climate change, animal and plant species had millions of years to genetically adapt to the gradual onset of climate change. If climate change nevertheless occurred quite suddenly, as seems to have been the case during the first cold periods at the beginning of the Pleistocene, species did indeed die out; in the case of the Douglas fir (genus *Pseudotsuga*) mentioned above, this seems to have been the case 1 to 2 million years ago. Human interventions tend to be very short-term compared to natural climate changes, occurring over the course of several decades to a few centuries. This is too little for plant and animal life to adapt genetically-evolutionarily to changing environmental conditions. A mass extinction of species is therefore, if not to be expected in every case, at least possible in the case of strong human impact, purely from the point of view of the interrelationships.

Tree Species Composition

The orientation of tree species composition to the natural forest community, as required by standards for the certification of sustainable forest management (e.g. FSC), is not expedient in view of the challenges of climate change. On the one hand, forest communities are losing natural tree species such as ash or elm due to pathogens, so that they become poorer in species and less adaptable. On the other hand, this orientation towards the previous site conditions offers no guarantee that the tree species naturally occurring at the site to date or in the past will also be the best adapted species in the future.

(Wissenschaftlicher Beirat für Waldpolitik 2016).

Author's note: This inconspicuous passage embedded in the expert report contains "a lot of explosive material" from the point of view of nature conservation: The ideal of the "natural" composition of a forest is fundamentally questioned. However, this is in no way intended to deny the competence of the "experts". This is just another example of frequent conflicts between very different objectives in nature conservation and environmental protection.

The effects of climate change on glacier areas and glacier budgets are also strongly discussed in the research literature and in the media. While Germany has only minimal **glacier areas**, which are rather extended firn fields of only a few hectares each and which are not expected to have a major impact on the landscape aesthetics or on the water balance (Schneeferner and Höllentalferner in the Bavarian Alps; Blaueis, Watzmanngletscher, Schöllhorneis and Eiskapelle in the Berchtesgaden Alps), and are also unlikely to survive the expected climate change in the coming decades due to their small size and low altitude, the central alpine glaciers are often still extensive (Fig. 33.6). The glaciers, even if their glacier tongues often reach deep down, as the "uppermost

Fig. 33.6 Central Alpine summit panorama. The transition from white firn snow to grey glacier ice is clearly visible on the valley glacier in the centre of the image. In late summer, this transition marks the snow line and thus the transition between the glacier's nutrient area (surplus of snow and ice throughout the year) and the glacier's consumption area. Glacier surfaces, although almost inanimate, represent important influencing factors in the high mountain aesthetics as well as in the water balance of the mountains. Especially the smaller alpine glaciers are threatened by climate change. (© Björn Bartzel)

storey" of the high mountains are deeply anchored in the consciousness of Austrians, Swiss and South Tyroleans. But here, too, climate warming will lead to a rapid reduction in the area of the large glaciers, but also to the complete disappearance of the small cirque glaciers in particular. It is often the large valley glaciers that react particularly sensitively to climate change due to their large mass balance at the tip of the glacier tongue and often retreat by several metres per year. The glacier areas, which are so important for the landscape aesthetics of the large mountain ranges of the Central Alps, are of course almost devoid of higher life, with the exception of a few specialists such as the glacier flea (*Desoria saltans*) among the arthropods. However, nature conservation in the extreme high mountains does not focus on species and habitats of species, as is otherwise the case, but rather on the impressive and here essentially abiotic nature as a scenic spectacle. Especially the conclusion with "eternal snow" makes the view from the deep valley zones filled with green forest and meadow vegetation all the way up to the top appear particularly attractive in contrast. Rarely as clearly as here in the rock and ice region does nature become tangible as a counter-world to the humanly used and often overused environment (Chaps. 1 and 2). In Austria, a large part of the eastern central Alpine main ridge is protected in the Hohe Tauern National Park. The western Austrian part in the area of the Ötztal Alps enjoys the less intrusive status of a nature park (explanation of both categories in Chap. 6).

So far, we have been talking about the effects of climate change on nature as an object of nature conservation. However, the feared and expected climate change also has another component, which results in increased social measures to combat it.

Germany, as one of the largest industrial and economic powers, imposed the **Renewable Energy Sources Act** (literally: Act on the Priority of Renewable Energies) on itself in 2000, which has since been amended and expanded several times (2003, 2009). This law sets compensation rates that favour so-called renewable energies over fossil or nuclear energy sources. Such a restructuring of the energy supply, which is primarily intended to prevent further warming of the earth's climate, would also meet the objectives of nature conservation. However, it is apparent that, beyond this common objective linking nature conservation and environmental protection, the expansion of **energy production from biomass** is decidedly detrimental to nature conservation, and must be seen as much more negative than the climate change it might prevent.

This is mainly due to the fact that intensive agriculture, which is mainly used for food production, is already **competing** for **land** with nature conservation. Now, in addition to nature conservation and food production, the production of energy from biomass is added to the competition for scarce land.

Since other industrialized countries besides Germany are also promoting energy production from biomass (in the USA, around 30% of the maize harvest (*Zea mays*) is now used for energy), an ever larger proportion of land is required for agriculture. Demand for **bioenergy** has now become the main cause of tropical forest destruction in both South America and Southeast Asia (Fig. 33.7). In the

Fig. 33.7 Peatland rainforest in Southeast Asia. This somewhat homogeneous-looking but species-rich forest type covered most of the lowlands of the Malay Peninsula, Sumatra and Borneo until about half a century ago. Since then, it has predominantly fallen victim to oil palm plantations. Palm oil is considered a renewable energy source and is also used on a large scale for European combined heat and power plants. It is precisely here that the problem of renewable raw materials becomes apparent. In the picture, the vegetation in the foreground is discoloured brown due to burning caused by clearing. (Own recording from the 1980s)

former case, forests are being cleared primarily for the benefit of the increasing cultivation of maize and sugar cane; in the latter case, for the benefit of palm oil plantations, which, among other things, are the main source of the biofuel E10 in Germany. The great importance of the use of grain harvests for energy purposes is reflected in the close correlation between the price curves of wheat and oil that has been observed for years.

It must be made quite clear: our earth is too small in terms of surface area to satisfy the various demands: namely to protect nature, especially tropical forests, *and* to adequately supply a still growing world population with food *and* to make a significant contribution to the global energy supply via bio-production.

Land Use for Energy Crop Cultivation

The amount of land used for the cultivation of energy crops has risen sharply in recent years and has a corresponding impact on nature and the landscape.

Current estimates assume a cultivation area of 2.2 million hectares, which corresponds to about 17.5% of the German arable land. Corn for biogas production is cultivated on about 0.9 million hectares, and rapeseed for biofuels on 0.6 million hectares. This means that more than every third hectare of maize and almost every second hectare of rape is used for bioenergy production.

(Natur und Landschaft 91, 2016, H. 12, p. 575).

In addition to the expansion of agricultural land for the production of biomass, the **intensification of agricultural production** due to additional demand must also be seen as negative for nature conservation goals. For example, grassland that is valuable in terms of nature conservation, which is after all home to several ground-nesting birds as well as a number of wild herbs, has increasingly been converted into maize fields in recent years, especially in Germany.

Using biomass derived from waste or from sewage sludge for energy supply, on the other hand, does not conflict with the objectives of nature conservation. However, the major breakthrough in renewable energy supply will not be achieved with this focus alone.

The demand for **renewable energies and raw materials** has now also reached the German forest. After decades of rather low timber revenues, which caused the economic interest in the forest to wane and made it possible to remove nature priority areas in the form of protected forests, for example, there now seems to be an energy turnaround in the forest. Supporters of more natural forests are currently fighting bitterly against people who, based on their own interests (large private forest owners), are calling for a turnaround in the forest and demanding intensified timber production. In the end, the fight for a national park in the northern Black Forest was also under this sign.

From an energetic point of view, the designation of a forest that has been left to its own devices is unproductive. The dying giant trees do not generate any surplus energy that can be used by humans. However, leaving trees to mature is also inefficient from an energy perspective, as a forest approaching maturity produces less annual wood increment than a youthful forest. So what to do? Cut down the forest as a kind of coppice every 10 or 20 years (Fig. 33.8)? For nature conservation, this is a horrifying idea, because most species of the forest would not be able to cope with this matted network of young tree growth and would disappear.

However, other forms of sustainable energy production are also more than questionable from a nature conservation perspective. These include, in particular, the use of rivers for **hydroelectric power generation.**

On the one hand, the water turbines, even when protected by nets, become death traps for countless fish. On

Fig. 33.8 Farmer's schematic row planting of fast-growing hybrid poplars, which are felled by a log harvester after about 20 years. Is this what the "energy forest" of the future will look like?. (Own recording)

the other hand, the river water must be dammed in almost every case in order to be able to regulate power generation at short notice by controlling the outflow according to need and demand and in order to achieve the necessary gradient for the power thrust of the turbines. Rivers are also dammed for reasons of navigability and flood protection (in seasonally dry regions also for the provision of service water).

The hydrological conditions of such a **reservoir** no longer correspond to those of a river, but rather to those of a standing body of water. For example, the oxygen content is reduced, since the water hardly moves any more and thus there is less oxygen input from the air. This results in a shift and also often impoverishment of the expected species spectrum. In addition, the natural ability to decompose organic substances in the sense of a self-purification of the water bodies is reduced with a reduced oxygen content.

In addition, dams prevent the migration of species. Several fish species, such as salmon (*Salmo salar*), eel (*Anguilla anguilla*), *allis* shad (*Alosa alosa*), and cyclostomes (lampreys) are distinct **migratory species**. However, even fish that are not considered migratory species often exhibit astonishing mobility, which serves to promote genetic mixing and thus genetic health. For example, it has only recently been possible to prove that the originally assumed three trout species, brown trout, lake trout (*Salmo trutta*) and sea trout (*Salmo trutta trutta*), are genetically almost identical and belong to a single species, the members of which must continuously interchange through migration. However, these migrations are prevented or severely restricted by barrages. The fish ladders prescribed as compensation are often not accepted by the migrating species or form bottlenecks for the migrations, where predators such as the grey heron are already waiting to prey on the fish.

As a result of river development, all migratory fish species in Germany have become rare or already extinct.

Conservationists Find Themselves in a Quandary

Coincidentally, exactly on the day after Tübingen Mayor Boris Palmer (Greens) criticized the priority of species protection over wind power, the Nature and Biodiversity Conservation Union Baden-Württemberg (Nabu) had invited three experts on the topic of kites for Friday afternoon. "We nature conservationists are just torn apart," said state chairman Andre Baumann on the sidelines of the meeting in Stuttgart. "On the one hand, we want the protection of the red kites, on the other hand, we want the energy turnaround." The Nabu wants to know where rotors and the endangered bird then now actually exclude each other. Baumann speaks of open questions.

It was exactly this debate that Boris Palmer had called for on Thursday. On the occasion of the rejection of one of the wind farms in Horb am Neckar because of a red kite occurrence, Palmer had mockingly said: "If a location for wind power is published, a kite is sighted the next day."

(Michael Petersen in Stuttgarter Zeitung, 17.8.2013).

Comment: The conflict as such seems indeed hardly resolvable if one does not assume, as is the case with the actors involved, a clear conceptual and substantive separation of nature conservation and environmental protection.

On the other hand: After the buzzard (*Buteo buteo*), the "bird threatened with extinction" is by far the most common hawk-like bird of prey in large parts of Baden-Württemberg. It is encountered almost daily on migration in suitable terrain, i.e. semi-open landscapes rich in structures, and almost all possible Kite territories appear to be occupied. However, Germany has a special responsibility for the conservation of red kites because about half of all red kites live in Germany, the other half mostly in neighbouring areas such as Alsace, Denmark or western Poland.

New Threat to the Red Kite

If several citizens' initiatives were not fighting against the project, EnBW's actions might not have become known: Back in June, the electricity company applied to the Stuttgart regional council for an exemption from the ban on killing red kites under species protection law in order to be able to build two wind turbines near Adelberg (Göppingen district). The turbines between Adelberg and Börtlingen are to be the highest of their kind in the region with a total height of 230 m above sea level—but are apparently located in the middle of a flight corridor regularly frequented by red kites. They could therefore easily become a death trap for the birds of prey.

(Karen Schnebeck in Stuttgarter Zeitung v. 23.8.2016).

Irritations Within the Association Between Nature Conservation and Environmental Protection

Hans-Josef Fell Leaves BUND

With "immediate effect", **Hans-Josef Fell** has cancelled his membership of the Bund für Umwelt und Naturschutz (BUND). In a five-page letter, **Fell explained** his decision by saying that the Bavarian branch of the environmental organization was endangering the nuclear phase-out and hindering climate protection. The step was not easy for him, as he "greatly appreciates the nature conservation work of the many volunteer members of BUND," said the Alliance Green politician. However, he "no longer wants to support the increasingly counterproductive energy policy work" of the association. Fell sat for his party in the Bundestag for 15 years, **but missed** re-entering parliament in **2013**. He is one of the **four authors of** the Renewable Energy Sources Act (**EEG**).

(https://www.klimaretter.info/umwelt/nachricht/19252-hans-josef-fell-verlaesst-den-bund; reported July 2015; posted Jan. 9, 2017).

Comment: One who has understood the tendential incompatibility of nature conservation and environmental protection.

Fighting Climate Change: Rewetting Peatlands as a Contribution to Protecting the Earth's Climate?

Only around 6% of the agricultural land in Germany is moorland. Nevertheless, with 41 million tonne CO_2 they are responsible for almost all (93%) greenhouse gas emissions from agriculturally used soils (Umweltbundesamt 2013 in Schäfer and Kowatsch 2015)—still approx. 4.4% of German greenhouse gas emissions (TEEB DE 2015).

[...]

If the figures are compared with other climate protection measures in the area of land use, it becomes clear how cost-effective the rewetting of peatlands is for reducing greenhouse gas emissions. The abatement costs here are only between €0 and €15 per tonne of CO_2 saved. In contrast, reduced greenhouse gas emissions through the use of hydropower, wind power or biomass incur significantly higher costs of €22–459 per tonne of CO_2.

(Schlegelmilch et al. 2018).

Comment: The "costs" of the rewetting of peatlands seem to be rather cautious here. According to classical market-oriented economics, subsidies are in themselves harmful, as they distort the market equilibrium and lead to cost burdens elsewhere, especially in the form of taxes. If one follows this established approach, the abandonment of agricultural use on peatland sites would even result in cost savings.

34

Hikers, Cyclists, Motorists: How Leisure Modalities Shape Our View of Nature

Driving *Looking into the middle distance, about 30 or 40 m, to where the car in front is driving or where new signs or road signs become legible. In between, the gaze repeatedly goes to the dashboard. Then again into the distance. Was there a tree on the side of the road? A bird of prey? Passed by. No matter, every close look is a traffic hazard. So keep looking straight ahead. In between, take a look at the dashboard and keep an eye on the direction of travel.*

Cycling. *The view about 10 or 20 m in front of the wheel. Watch the path, there could be potholes. A tree on the side of the road, a bird of prey on it. A quick look. Could have been a buzzard. There's not enough time to take a closer look, or even to observe. Dismounting is uncomfortable. Besides, the bird of prey would then probably seek the distance anyway. So we drive on. The landscape in view, distant hills on the horizon, a group of trees gradually approaching. But no view for the detail.*

K.-D. Hupke, *Nature Conservation*,
https://doi.org/10.1007/978-3-662-66159-8_34

Hiking. *Sometimes tiring slowness. A wide treeless plain without detailed impressions. Boredom. Time to think. Or for conversations. Or to look closely. A tree on the side of the road, a bird of prey on it. Stop, look closely, observe. Register details. To memorize. The bird of prey rises in wide wing beats. Look after it. Register the flight pattern. Afterwards look at the ground again. A small flower blooms at the wayside. Never seen before. Standing still. Bending down. Breathe in the fragrance.*

The lines just formulated are intended to suggestively illustrate our experience of the outside world with different modes of locomotion. After all, the preferred modes of locomotion have changed over the past decades. This alone, and not just mass media influence, has caused a revolution in our **perception of the environment**. Basically, only those who move slowly, usually on foot, have anything to gain from the species and forms of nature. Likewise, only the pedestrian experiences the full extent of the monotony of our cultivated and machine-oriented agricultural landscapes, while the cyclist has already quickly passed through the most monotonous large-scale agricultural areas.

The speed and fleetingness of modern means of transport also corresponds to the fleetingness of the optical-aesthetic animation of our television worlds. In comparison, experienced nature outdoors is often less spectacular at first glance, requires background knowledge for full enjoyment and still has to do without the explanatory voice from the background. This makes it a problem for many contemporaries, resulting in experienced boredom.

But at this point we should not list another example of cultural pessimism. Of course, nature is an offering to body and mind. On the other hand, one can also be quite happy without a special sensitivity **for experiencing nature**. After all, there are also music philistines or the historically ignorant who do not necessarily suffer from this “deficit”.

However, people who completely renounce knowledge of nature and observation of nature are also renouncing an area that counteracts the other horizons of experience of the inner life of society and thus represents a kind of counter-world. Of course, within the framework of a nature shaped by human beings—and where else would we have another one—we in turn encounter social reality. Systems of nature, as nature conservation understands them, however, always mean a degree of detachment from human and social influences. When it comes to shaping nature, man is always just one factor among several otherwise natural influences. All in all, and considering that we are dealing with the interaction of independent living beings, the natural factors in this context are at least as important as the human influences. Where else in society can one find such a partially autonomous sphere?

Apart from that, I would like to venture a thesis that is provocative for many: Nature does not need man, they say. But man does not necessarily need nature either; at least not the idyllic nature oriented towards the image of the traditional agricultural landscape, which is so favoured by nature conservation in Europe. Human society would survive even without nature conservation. Humans have proven to be incredibly adaptable in their tribal history as well as in the present; more so than any other biological species. He doesn't need conservation areas either. Even if the climate were to change (even slightly) as a result. Human societies can survive in many climates—not to say almost all.

It would nevertheless be regrettable if the view of an endangered nature were to disappear from the public consciousness.

35

Man and Nature: A Constructed Opposition?

People of Asian origin are often accused by Europeans of being in need of harmony and of avoiding conflicts. This is confirmed in many encounters between Europeans and Asians. The only question is where this circumstance comes from.

Even though most of us today are no longer explicitly religious, our thinking is nevertheless strongly shaped by thought mechanisms that have been inherited from religious ideas. In the Western religions of Judaism, Christianity and Islam, which are based on the Old Testament traditions, the conception of the world is very much shaped by the fundamental opposition between the Godhead and the counterworld of the devil. Man has to make a choice. A middle way or some kind of compromise is out of the question. Any wavering on the right path is automatically of the devil.

From this basic conception, even in secular life, it is difficult for a differentiated view of morality to prevail. We

K.-D. Hupke, *Nature Conservation*,
https://doi.org/10.1007/978-3-662-66159-8_35

still widely copy this bipolar pattern today and tend to see the whole world in bipolar dichotomies: Government and opposition, employer and employee, rich and poor, man and woman. And of course: nature and man.

The Far Eastern Asian religions on the other hand, do not have a monotheistic deity. They have many gods that coexist and can be worshipped and adored in parallel. One and the same deity can show himself in different states, so-called incarnations. Such religions educate tolerance. Tolerance suggests compromise and middle paths.

Let us go into one of the main **oppositions** of modern Western thought and the content of our book: the **opposition of nature and man.** Often, human influence on nature is experienced as sinister, in the spirit of a moralizing religious preconception. Many conservationists have spent their lives in search of a **nature untouched** by man, which they are unlikely to find today, even in high mountains and tropical rainforests. This seems downright tragic (Fig. 35.1).

Fig. 35.1 The strict model of separation of nature and culture does not apply: Here, the desired "natural conditions" are helped along in a nature reserve by dredging. (© Gerd Markert)

Wild plants and animals and the communities or biotopes they comprise are exposed to a variety of different **environmental conditions.** Few of these are merely favourable; many are detrimental or require certain adaptive processes to survive. Many of these environmental conditions are natural, such as geological subsoil, soils, relief and climate. To a greater or lesser extent, there is also a varying degree of **human influence** almost everywhere. However, in general and in principle, this influence is not so different from the other factors that are perceived as natural. Human influences also modify biotic communities and biotopes, but as a rule they do not destroy them any less than, for example, drought in the deserts or cold in the polar regions would do. And yet there are so many people who rave about the beauty of nature in the desert or the pristine nature of the polar regions: although in both cases these are rather species-poor habitats under extreme environmental stress, thus not dissimilar to many anthropogenically impoverished habitats.

The search for untouched nature reveals more about the flight from civilization and the idealization of conditions far removed from civilization than about nature itself.

On the other hand, man is also nature. He is this with his physicality, which is by no means controlled by civilization, even if some modern physicians, biotechnologists or pharmacologists should have something like this in mind. He is the provisional end point of an evolution lasting millions of years, which is completely nature from beginning to end. Even if he behaves as if he were far from nature, man can no more escape nature than the Old Testament Jonah could escape his deity.

Even the mental potential of man, which is conspicuous in the context of the rest of nature, is nature in so far as it is based on a eutrophically developed organ of the human body, and this brain of man has also developed in the course of evolution by processes which we can only see as nature.

Nevertheless, the self-reflecting human being who determines his spiritual position has every reason to think about the fact that and why he wants to preserve and protect the diversity that has come down to us. He is the only part of nature that can question himself and his actions. For the first time in hundreds of millions of years, something like responsibility arises in man: for himself, but also for the nature surrounding him. This makes the extinction of species by humans different from a comparable natural process, for example at the beginning of the cold ages or in another natural climate change.

36

Search for Ideas: How Can Nature Conservation Be Justified and Anchored in Society?

After what has been said in the previous chapter, there may not be a fundamental duty, but there are good reasons to protect nature (whatever that may mean in detail). In a world dominated by media-driven discourse, nature conservation is urgently dependent on **positive publicity.** Unfortunately, this is rarely done, or often done in a highly unprofessional manner. The present chapter is intended to introduce some possible solutions into the discussion.

The poor **public image of nature conservation** begins at the edge of a nature reserve. This is usually marked by a sign with the familiar symbol (in western Germany a bald eagle, in the east an owl). Below this often follows a list of prohibitions: Camping or making fires, for example (Fig. 36.1). The great opportunity to explain to the hiker or walker what is special and worth protecting about the designated nature reserve (Fig. 36.2) is usually missed.

The same applies to the unconditional ban on entering the areas off the paths. How is an extensive object such as a

K.-D. Hupke, *Nature Conservation*,
https://doi.org/10.1007/978-3-662-66159-8_36

Fig. 36.1 Panel with list of bans. (Own recording)

nature reserve to be experienced if not by walking on it? The argument that trespassing causes damage is only valid to a very limited extent. After all, "trespassing" by sheep, for

Fig. 36.2 Pictorial panels with explanations, which present characteristic flowering herbs and butterflies on the left, and give an insight into ecology and landscape history on the right. All depicted conditions and individual species are exactly aligned to the situation of precisely this site in a rough grassland area. (Own recording)

example, with their sharp hooves, causes at least the same damage during a maintenance operation as trespassing by humans; not to mention a pack of wild boar. In addition, a massive trespassing by people interested in nature is not to be expected anyway, due to the lack of broad public interest. The argument of disturbing animals during breeding and rearing does not seem to hold water either. A lacing fox, and there are more than enough of them, is experienced as at least as threatening by a ground-nesting bird. Here again the attempt to exclude humans from nature shines through, as quasi the only pest nature knows. If the breeding success of the capercaillie is endangered by treading on the edge of the upland moor, the relevant areas should definitely be extended in order to make capercaillie and humans compatible in the area. Apart from the fact that the access qualities of a moor pine forest are in any case not such that, as said, a stronger rush of nature enthusiasts could be expected.

Similarly, the catching of individual nature-protected animal "objects" such as tadpoles can be viewed quite loosely in individual cases. There is probably no better form of education to become a nature lover and later a conservationist than observing self-caught tadpoles in one's own mini-aquarium. However, since all native amphibians are now protected, even biology teachers and other multipliers are forbidden to do this. This regulation of the Nature Conservation Act applies in principle and even to a removal from wagon tracks, which should dry out already after a few rainless days and become a grave for their inhabitants with high probability.

Nature conservationists and their legal arm should recognize that humans, as soon as they enter the nature reserve to observe or even catch an animal, play no other role than the (other) animal visitors or inhabitants. In this case, nature conservation must ensure that the desired **ecological qualities of the protected habitat are** maintained, whether through maintenance or process protection. Whether single animal or plant individuals are removed or damaged in some other way is usually unimportant. Individuals of species that are actually worthy of protection are actually always damaged in some way, already by any maintenance intervention. Conservationists should not look at nature reserves like a priest looks at the holy of holies of his church, which must not be touched under any circumstances. For the holy of holies of a church or temple is, if you will pardon the expression, something museum-like and somehow dead. A relic or shrine does not change except through general decay over time. By contrast, in a semi-natural habitat, there is always some kind of birth and death. The latter follows from the very fact that most species produce more offspring than can by any stretch of the imagination survive. Therefore, the death of an individual (not of a species) does not necessarily affect the conservationist (unlike the animal rights activist), as long as the properties of the protected habitat remain intact.

In view of these considerations, a catalogue of demands for a modern and socially integrated **nature conservation that** is **sustainable in** this sense would include the following points:

1. Nature reserves must become aware of their **broad educational task**, but this is only happening very gradually. There should no longer be a nature reserve without motivating and meaningful picture boards with short texts that highlight what is special about the area at hand (but thus also: put it up for discussion).
2. One key question always arises: **For whom should nature be protected?** Who should it serve, who should have direct access, who should be allowed to look at it?
3. Demand that **nature conservation areas be opened up** to an interested public. Where damage is to be expected due to an excessive density of visitors (e.g. in the Lüneburg Heath as a local recreation area for Hamburg), the corresponding protected areas must be extended in accordance with demand until the pressure of use has settled down to a tolerable level. Humans should have no less rights to nature than, say, foxes *(Vulpes vulpes)* and wild boar *(Sus scrofa)*.
4. **Open discourse** is always preferable to the imposition of nature conservation rules. The discussion about which nature should be protected in which way should be conducted with the involvement of all interested parties and a broad public. From what has been said so far, it is clear that in the field of nature conservation there can be no such thing as a designated expert. Biologists and land-

> scape painters, recreationists and local historians, young couples in love and children at play are all equally experts in their own right who seek out and use nature. It is precisely the diversity of approaches that makes a discourse on nature interesting in the first place and does justice to the many connotations of nature that have arisen historically.

Exclusive access to nature is more than questionable in a society that otherwise defines itself as open. Moreover, by excluding people, especially young people, a nature education opportunity is missed.

37

Nature Conservation Versus Zeitgeist?

I do not want to glorify everything that was part of life in my own youth. But one thing seems certain: school **biology lessons** were more oriented towards the tangible nature of species and habitats than they are in the present day. I remember two thick books at the beginning of high school: the Pflanzenkunde-Schmeil and the Tierkunde-Schmeil. Both were closely oriented to biological systematics, which they taught students in what I think was an impressive way: in an assortment of species and life forms that were vividly described and backed with color pictures. The pictures usually depicted not so much the individual species as a section of a biological habitat. These heavy books contained far too much content to be worked through completely at school. But they encouraged the students to do their own reading at home. That is quite remarkable for a school book.

The school's special introduction to **species knowledge and biological systematics** – even if, like so much that is taught in school, it did not leave very deep traces – nevertheless

K.-D. Hupke, *Nature Conservation*,
https://doi.org/10.1007/978-3-662-66159-8_37

produced a certain willingness to get involved with nature outside of school. This happened during the many hikes that were still a widespread part of family life at the time. A good help was provided by identification booksIn the German-speaking world, these mostly came from the Kosmos publishing house and, in addition to a popular title *(Was blüht denn da?, Was fliegt denn da?), were* distinguished by excellent colour prints of the species concerned.

However, the awakening of interest in individual species at school also created a ground for the acceptance of animal programmes on television, which was still young at the time. The Grzimeks and Schuhmachers, the Sielmanns and finally Horst Stern (Chap. 3) created and deepened an interest in animal forms that gradually turned more towards the "ecological" in the 1970s.

Of course, this **socialisation** towards the knowledge of animal and possibly plant species, as has just been described, also has something very subjective about it, which probably applies to my person in particular. But a particular kind of socialization most seldom proceeds alone. This pattern was important enough to make conservation more **acceptable in society**. The monthly magazine, *The Little Animal Lover,* was probably sold with success in almost every school; a magazine with a mixture of conservationist and animal protectionist ambitions. Hundreds of thousands of girls and boys became involved with an astonishingly differentiated view of animal species, their way of life and their behaviour.

In the meantime, the broad presentation of individual species and life forms has all but disappeared from school biology lessons. Instead, the exemplary, the natural law-like and the rule-like have prevailed. For good reason, since the **development of biology** as an underlying natural science (Table 37.1) had also continued. Instead of a collection of what occurs in nature, laboratory methods and a technically oriented view became dominant. Some claim that

Table 37.1 Biology as an important reference science for *nature* conservation between *natural history* and *natural science*

Characteristic category	Biology as natural history	Biology as a natural science
Reference to the individual species	Each species is important and in principle part of the scientific interest	Focus on character types that include "all essentials".
Way of working/ thinking	Descriptive individualizing	Natural law vicarious-exemplary
Performing large form	Description, lithography	Experiment
Research and presentation method	Qualifying	Quantifying
Rough chronological order	Traditional	Modern
Forecast	Fading	Increasingly
Related sciences	Arts and other cultural studies, historical and human sciences	Exact sciences (chemistry, physics)

Source: Own representation

biology has only now become a true natural science. In any case, in line with the academic orientation science, plant and animal systematics as well as the multitude of species and forms presented largely disappeared from biology lessons at school from the 1970s onwards. Today's school leavers have virtually no knowledge of species from school any more. Nature conservation is thus losing the knowledge base on which it has always been able to build. In addition to the still high verbal acceptance of the concept of nature conservation, there is also a widespread and increasing alienation from **the content of nature conservation.** This is also reflected in the structure of nature conservation organisations. Either they are aging to the point of survivability, or they are adapting to the spirit of the times by adopting the goals of animal protection and general environmental protection, which, due to their lack of ability to differentiate, few members should notice anyway.

The fact that jogging and cycling (Chap. 34) have largely replaced hiking and walking as outdoor exercise activities means that, on another level, the opportunity for in-depth observation of wild animals and plants is declining. The declining connection to agricultural and horticultural design on farms and in private gardens, as can be widely observed, goes in a similar direction.

As a result of modern lifestyles, nature conservation is increasingly losing **access to nature**. As a result, young conservationists are becoming scarce, while the older generation of conservationists tends to have moved into nursing homes or passed away. All the more reason, therefore, to place special emphasis on educational work.

38

The Benefits of Diversity: Reality, Poetry or Esotericism?

In science, politics and the media, the **benefit of nature in** the sense of the objects of nature conservation, i.e. the diversity of species and habitats, is simply assumed to be largely unquestioned. Attempts to justify and derive such a benefit are more rarely found.

This argumentation of benefit is already difficult because it can have many facets beyond a superficial economic gain. Above all, we need to think of added value linked to interests, such as the interests of users of the landscape, from farmers to people seeking recreation, as well as the interests of nature conservationists themselves.

Nevertheless, every discourse in a reasonably open society is based at least on the attempt to convince the other side by means of arguments. Arguments need a rational basis, but one that does not have to forego either an emotional component or subjective concern.

In this context, reference is always made to the dependence of the individual and of human society on natural

K.-D. Hupke, *Nature Conservation*,
https://doi.org/10.1007/978-3-662-66159-8_38

products. All our **crops and livestock**, however remotely they are produced today, are descended from **wild forms.** Not everyone would go as far as the former president of the German Federal Agency for Nature Conservation, Hartmut Vogtmann, who also sees the protection of the diversity of old crop varieties as the primary task of nature conservation. Nevertheless, wild plants and wild animals form a pool from which new crop varieties can in principle be bred.

However, existing livestock and crops, often genetically impoverished by breeding processes, can also benefit from the existence of wild forms with which it is possible to cross. Repeatedly, the crossbreeding of wild forms has made it possible to increase the vitality of cultivated forms, or cultivated forms have thus developed the potential to respond to new market requirements.

Wild plants are important not only as food, but also for the **development of new medicines.** Most medicines in use today have a natural counterpart, mostly in natural plants, which in many ways can be seen as a rich chemical laboratory (or a well-stocked pharmacy). The list of important medicines with a natural source is almost endless, ranging from the painkiller aspirin (the natural substance is found in native willow bark) to the modern chemotherapeutics and cytostatics used in cancer therapy, such as vincristine and vinblastine (from the Madagascar native *Catharanthus roseus*).

Why substances produced in and by nature are more likely to work as medicines than purely synthetically produced substances is briefly explained using the example of **antibiotics.**

The most important antibiotic, at least historically, is penicillin, which was discovered almost a century ago by the British physician and scientist Alexander Fleming (although it did not come into use until many years later). It is a natural component of various moulds. Like penicillin, almost all antibiotics used today are derived from natural substances (but are usually produced synthetically).

However, not only humans have problems with **bacteria** (and only against these antibiotics can be used by definition), but perhaps even more so amphibians living in the tropics, such as the numerous South and Central American tree climbing frogs (Dendrobatidae). Their bodies are not covered by a skin with a firm horny layer as in humans and other higher vertebrates, but by a mucous membrane. This offers little mechanical protection against bacteria, but due to its richness in nutrient salts, which are rare in the tropical rainforest, and the perpetual warm and humid environment, it is an ideal breeding ground for bacteria. This means, however, that in order to survive at all, tropical amphibians had to "come up with something" in evolutionary terms to avoid becoming victims of skin-colonizing pathogenic bacteria. They have had several tens of millions of years to do this and have developed numerous natural antibiotics. In a certain sense, antibiotics research is treading on pre-explored territory when it searches for new antibiotic substances on amphibian skin swabs. This makes the search for new antibiotics easier than if new synthetic substances were to be developed indiscriminately and tested for their efficacy as antibiotics while at the same time being harmless to patients. The interest of pharmaceutical companies in tropical biodiversity is also nourished by such considerations.

At the same time, however, we must warn against seeing nature only as a **gene pool for** the benefit of agriculture and pharmaceuticals. Nature can be just as destructive. The origins of major recent **infectious diseases** such as AIDS and Ebola lie in tropical rainforests. The world's greatest potential for viruses, bacteria, protozoa and fungi resides in tropical moist forests. It is almost a stroke of luck that although AIDS and Ebola are usually fatal if untreated, they are more difficult to transmit than many flu-like infections. The latter, on the other hand, are usually comparatively harmless. However, it is conceivable and probably feasible in the long term to have an absolutely lethal pathogen which is as infectious as a mild summer flu. Never before in human history have the conditions for a pathogen been so favourable. On the one hand, they draw on a potential of now more than 7 billion people, whereby in larger populations the probability is generally higher that a pathogen will emerge or jump over. On the other hand, we are so mobile today that any such pathogen would probably spread worldwide within a few weeks, as has also been shown with the comparatively "harmless" COVID-19. However, we cannot retreat into our homes and shut ourselves off from the outside world for months before an infection, as the inhabitants of remote farmhouses were able to do in plague times. After all, our needs have become far too diverse and even the supply technology would collapse within a few days if no one left the house for fear of infection.

This horror scenario is intended to make it clear that tropical forests not only have the potential to provide for humanity in the future. They have just as much potential, if not for their destruction, then for their decimation. It is characteristic of public relations work from NGOs to the media and politics that only the former is perceived.

But maybe that's for the best.

The considerations so far suggest that nature conservation should be seen as a kind of biotope protection for a construct. Finally, it should have become clear that we are dealing with problem approaches that can only be derived from the history of our relationship to nature and landscape and from their idealization.

If one criticizes the substantive and formal weaknesses in the chains of argumentation in favor of the previous and strongly official, but also NGO, nature conservation, as is the case in a large part of the chapters of this book, one must also include the alternatives in this line of thinking.

If nature conservation did not exist, a strongly market-driven automatism would consume even the last patches of biodiversity, especially in rural, i.e. predominantly agriculturally used areas. It is possible that the inhabitants of these areas, in their vast majority, would hardly notice this and that only an urban minority with corresponding leisure interests would be affected.

On the other hand, areas of high biodiversity are almost always also areas of high aesthetics and great optical-visual richness. This is certainly true for traditionally shaped agricultural areas as well as for "primeval forest" protected forests. However, **diversity and aesthetics** are central categories for **quality of life,** even where they are not consciously perceived. They complement our sober economic as well as our scientific-technical rational world. Money and technical efficiency, however, cannot generate values *per se*, since they lack the upstream meaning that is *a priori* inherent in the Kantian sense. Economic and scientific efficiency only make sense if one knows what one needs this additionally created money for, what technology one wants to use for what purpose. However, goals and purposes can neither be defined by economics nor by natural science and technology. Rather, these goals arise from historical tradition, from personal experience and life practice, or from religious or philosophical discussions. Above all, however, goals and meaning are always something very individual, both in relation to individuals and to social groups, cultures or epochs.

This book can therefore be seen as a plea in favour of nature conservation, as soon as it clearly formulates its criteria, such as **process protection** or **protection of the traditional agricultural landscape,** or **aesthetic criteria.** Then it can be discussed. This reflection on its own foundations is always to be recommended to nature conservation, instead of labelling all efforts of official and private nature conservationists as a blanket advocacy of nature, as is so often the case. This label hides more than it explains.

In any case, society would be much poorer without nature conservation; perhaps less in the sense of biodiversity or nature, but rather for an impulse that builds on traditional values. This also contrasts pleasantly with the efficiency thinking of modern economic and scientific technocracies, whose goals and values are often even less clear than those of nature conservationists anyway.

In this sense, a society without conservationists would certainly be poorer.

Threatened Coffee Bean

Hardly anyone thinks of their morning cup of coffee when scientists bemoan the loss of biodiversity. But botanist Aaron Davis and his colleagues at Kew Royal Botanical Gardens in Sussex, England, do see a connection: worldwide, the two species Arabica and Robusta provide almost the entire harvest of the coveted beans for the luxury food. But climate change could cause problems for coffee cultivation, for example through more frequent periods of drought. Newly emerging pests or plant diseases could also massively decimate the harvest. However, breeders could solve this problem by crossing with resistant wild coffee species. With 124 species worldwide, there is plenty of choice. But 60% of these species are now on the Red Lists of the World Conservation Union IUCN and are considered endangered or threatened with extinction, the British researchers report in the journal "Science Advances". (....)

In order to preserve the biodiversity of coffee plants, researchers led by Aaron Davis are therefore proposing protected areas (Roland Knauer, Stuttgarter Zeitung v. 19.1.2019).

39

On the Future of Nature Conservation

Among other things, the previous chapters have tried to show that nature conservation is, in a certain sense, losing its conservationists. This chapter now attempts to make clear that, in addition, the nature to be protected is also subject to a problematic decline, which the nature conservation that has existed up to now can at best delay.

This **decline of** near-natural, or at least diverse, ecosystems is a process that began in excessive form in the nineteenth century with the **spread of industry** over large parts of Europe and North America. In ever new waves, other regions of the world are being included in this dynamic. At the moment, East Asia and Latin America are in a process of strong economic-industrial development, South Asia is about to follow. Presumably it will be Africa's turn after that. The process of spreading industrialization is coupled with an increasing hunger for energy and raw materials. Land, in itself a natural object with no actual production costs, is capitalised under these conditions and acquires an

K.-D. Hupke, *Nature Conservation*,
https://doi.org/10.1007/978-3-662-66159-8_39

economic significance and market value. Traditional land use systems worldwide are undergoing rapid change towards profit maximisation, the emergence of large-scale farm structures and acquisition by internationally active investors **(land grabbing).**

The development depicted is inconceivable without a **bourgeois society.** It is the actual motor of this development and even as a less biased observer one often does not know whether one should admire more its scientific-technical innovative power or despise its devastating effect on traditional systems of nature and culture. In any case, the bourgeois way of thinking that spread worldwide from the eighteenth/nineteenth century onwards, oriented not towards traditions but towards economic gains and individual advantage, and which is also reflected and reinforced in state legislatures and constitutions, leaves no stone unturned in any respect. Whether human labour or mineral resources, the wood of natural forests or crops: this society can really use everything in almost unlimited quantities, transfer it into the production process and convert it into dollars or euros.

One could argue that the **destruction of nature** already existed in pre-industrial times and in non-bourgeois social systems. That is certainly true. But the destruction of nature, for example in the late Pleistocene overkill, took place in millennia, and thus in a different temporal dimension. This is also true of other historical social interventions up to the early modern period. What is special about the current destruction of nature is its speed and its worldwide validity.

Even the impairments of nature in the former Soviet Union (a society explicitly defined as non-bourgeois), such as the drying up of the Aral Sea, cannot be seen in isolation from the embedding of the socialist system in the global economy. The states of the Eastern Economic Community in particular were involved in a competitive struggle for greater industrial commodity production efficiency; a struggle which they ultimately lost and which was the main reason for the end of the Soviet Union and the alliance it led.

But let us return to the very concrete threats to the survival of near-natural systems. The greatest threat to traditional to near-natural ecosystems, which are important to nature conservation, comes from an **intensification of agricultural and forestry use.** Such expansion and intensification of land use has occurred almost since time immemorial. Until about the end of the twentieth century, it was strongly motivated by a **growing world population.** The population of the earth is still growing at the beginning of the twenty-first century; however, the growth curve is flattening somewhat and the statisticians of the United Nations Population Fund (UNFPA) expect the world population to stagnate around the middle of the present century, and even to decline somewhat thereafter. So from this point of view, one could gradually give a slight all-clear.

On the other hand, of course, living standards in many emerging countries are also rising, and with them the demand for animal products and vegetables, which is further fuelling land consumption.

More decisive, however, is the increasing importance of agricultural products and agricultural land for the **production of renewable energy resources.** Already now, a larger double-digit percentage of grain (wheat, corn) is used for energy and is found, among other things, as an additive in fuels. The demand for energy, especially in industrialised countries and many emerging economies, is inexhaustible, and covering this demand from other sources appears to be problematic or impossible in the near future. This, however, will increase the demand for agricultural and forest land enormously beyond the needs of food production.

Basically, then, there are three requirements for limited arable suitable land, namely:

- To provide food for a still growing humanity;
- to make increasingly renewable energy raw materials available;
- as well as to enable near-natural ecosystems such as tropical humid forests.

Basically, if we included all these needs, we would need another second to third planet.

The increasing demand for agricultural and forestry raw materials has led to an **expansion of the areas used.** Thus, the hitherto almost unused wet and dry forests and savannahs of central Brazil and Paraguay are increasingly interspersed with islands of use, which in turn are becoming denser and the previously dominant near-natural landscape types are becoming increasingly isolated.

Similarly drastic, although often not immediately apparent, is a trend towards **intensification of land already in use. In** Central Europe in particular, grassland, which was previously very widespread and dominant in the rain-fed landscapes of north-western Germany and the low mountain ranges, is increasingly being broken up and converted into maize fields, for example.

Although intensively used grassland is by no means the ideal landscape for nature conservation, a meadow or pasture provides a habitat for perhaps two dozen species of wild plants and can be a nesting site for ground-nesting birds such as the skylark and lapwing *(Vanellus vanellus)* or a feeding ground for the white stork *(Ciconia ciconia).* A maize field offers none of this.

The renaissance of **wood as an energy raw material** should also be viewed with scepticism in this context. Decades of mostly low timber prices bordering on profitability have increased the acceptance of near-natural forest images with high proportions of unharvested old trees and deadwood, which are vital for many species (Chap. 14).

Now the forest owners and their associations are once again engaged in a very heated discussion on forest utilization. The main reason for this is the increase in the price of timber and firewood, which has been observed to a lesser extent for some years but is expected to be strong in the future, and which is due both to high energy prices and to the Renewable Energy Sources Act (EEG) in Germany. Wood has again become an important energy raw material. However, maximising the use of wood is contrary to the objectives of nature conservation in the forest.

If natural habitats decline, replacement strategies can be used to try to counteract this. The protection of near-natural habitats that are as extensive as possible as nature conservation in the narrower sense has been described in previous chapters. Other possibilities lie in the protection of individual species in a technical-artificial environment. Here, **zoological gardens** play a more important role in the conservation of large animals in particular. In fact, zoos have decisively redefined their role. Whereas zoological gardens saw themselves primarily as popular educational institutions until the 1980s, this interest is now subordinated to **breeding programs for endangered species** (Fig. 39.1). For example, subspecies of big cats such as the Persian leopard or the Sumatran tiger *(Panthera tigris sumatrae)* are now kept alive genetically through zoo breeding programmes because wild populations are increasingly too small and spatially too isolated to be able to do so. Release and capture programmes are designed to ensure that populations in the wild and in zoos worldwide continue to mix genetically and thus remain healthy and vital. Zoos have successfully countered criticism that they are complicit in the extinction of rare species in the wild through wild capture. In fact, the trend in zoos today is to keep far fewer animal species, but to embed them more strongly in a habitat modelled on nature and to have them reproduce there.

Fig. 39.1 The Syrian brown bear is a subspecies of brown bear *(Ursus arctos)* with light-coloured fur that is threatened with extinction in the wild, but will survive in zoos (as here in Heidelberg). (Own recording)

The conservation of plant species is easier to achieve in **botanical gardens.** Like zoos, large botanical gardens are often involved in species conservation programmes. For example, the Botanical Garden of the University of Heidelberg holds a total of about 10,000 plant species; this is three times more species than live in the wild in Germany and about 3% of all plant species worldwide in general. The Bromeliaceae collection alone (Fig. 39.2) comprises several hundred species, many of which may be considered endangered in their natural habitats in branch forks of South and Central American rainforest trees.

But is that what conservation really wants? To ensure the survival of species in gardens or even in genome banks, far from their natural habitats? After all, the survival of a species far removed from nature in itself offers something like a potential that at least does not exclude a later reintroduction at a site restored close to nature.

Fig. 39.2 The collections of pineapple plants (Bromeliaceae; in the picture) and orchids of the Botanical Garden of Heidelberg University preserve thousands of species, some of which are endangered or even extinct in the wild. (Own recording)

Sensational Success for Turtle Protection

There is great joy at the *Allwetterzoo*: For the first time, turtles hatched in Münster can be brought back to their original home! These are 14 Annam terrapins hatched under the care of Elmar Meier. The world-renowned expert is considered one of the most successful breeders of Asian tortoises. In the *International Centre for Turtle Conservation* (IZS) at the *Allwetterzoo,* Elmar Meier has been caring for various highly endangered species, some of which are extinct in the wild, with great commitment and on a voluntary basis since 2003.

On Monday, 12 August, Elmar Meier brought the 14 Münster tortoises to Rotterdam. Together with 57 conspecifics from Rotterdam Zoo and the stock of a German private breeder, they travelled to Vietnam. There, the reptiles are to ensure the genetic replenishment of the population. The Annam stream turtle *(Mauremys annamensis)* is one of Vietnam's endangered freshwater turtles. The population of this endemic species declined rapidly due to illegal trade and habitat loss. Since the 1930s, the species was considered lost and was only sighted again [in the wild] in 2006.

[...]

With a carapace length of up to 30 cm, the Annam stream turtle is one of the largest Asian pond turtle species. Today it lives endemically in a small area in the province of Quang Nam in central Vietnam. Formerly quite common and more widespread, the species was considered extremely rare by the middle of the twentieth century, as no more free-living specimens could be found. However, single specimens were offered again and again on Chinese markets and even occasionally turtles of this species were imported to Europe and the USA.

After a wild population was discovered in 2006, several breeding projects were initiated in Vietnam and China. In Germany, the *International Centre for Turtle Conservation* at the *Allwetterzoo Münster has* recorded good breeding successes with the Annam's brook turtle on the basis of sporadically imported animals and also works closely with the Rotterdam Zoo (Minor 2013, p. 11f).

Last Chance for the Pachyderms

Species conservation: The northern white rhinoceros is effectively extinct. But it is to be saved.

(...)

Even though three mighty animals are still trotting through the sanctuary, the end of the northern white rhinoceros is approaching. If it weren't for researchers like Joseph Saragusty, Thomas Hildebrandt, Oliver Ryder and a whole host of other specialists still grasping at straws: fertilization in a test tube. For this, in addition to concentrated expertise, only two things are needed to begin with: sperm cells and egg cells of the highly endangered subspecies. "We store the sperm of four males in liquid nitrogen at -196 °C," explains Joseph Saragusty. This keeps the frozen cells fit, and the researcher is very confident that they will do their job once thawed. In any case, this works excellently with horses – and they belong, along with tapirs, to the closest relatives of rhinos.

Roland Knauer in StZ v. 18.7.2016.

40

Follow-Up: Nature Conservation Expertise Is Everyone's Business!

In Olaf Kühne's excellent monograph *Landscape Theory and Landscape Practice* (2013), there is the following passage, which, although related to landscape science, explicitly includes nature conservation. One only has to replace "landscape" by "nature conservation" and the text fits our topic almost perfectly:

> *Alongside amateurish understandings of landscape* [understandings of nature conservation], *expert-like understandings of landscape* [understandings of nature conservation] *have developed with social modernisation: In the search for "solutions to problems", these are assigned to "professionally organised specialists". Expert landscape socialisation* [nature conservation socialisation] *usually takes place in adulthood in the form of a systematic imparting of landscape-related* [nature conservation-related] *patterns of interpretation and evaluation mostly in a landscape-related* [nature conservation-related] *specialist course of study (landscape planning and landscape architecture, landscape management, nature conservation,*

K.-D. Hupke, *Nature Conservation*,
https://doi.org/10.1007/978-3-662-66159-8_40

> *urban planning, geography, etc.) or a landscape-related* [nature conservation-related] *apprenticeship (e.g. landscape builder). This socialisation into a professional discipline mostly means the recognition and maintenance of its discursive rules (after all, it is necessary to defend oneself against the devaluation of one's own cultural capital).*—(Kühne 2013, p. 208).

The quotation illustrates the discourse dependency of professional constructs, to which nature conservation can also be effortlessly assigned. The self-interests are emphasized ("against the devaluation of one's own cultural capital"), without explicitly attacking the corresponding professional patterns. However, the way in which they are generated and communicated makes them appear at least perspective, if not questionable.

In principle, however, conservation issues are different from detailed biological research, for example, which main food plants the field hamster prefers or how to imagine the synthesis of DNA polymerase. The last two questions will (have to) be left to the experts.

Nature conservation, on the other hand, includes values and perspectives that stem from our everyday knowledge. This everyday perspective is not lost in nature conservation professionalization, but it is overlaid. The professionalized conservationist is indeed superior to the "layman" in terms of historical contexts and the view of the multi-perspectivity of conservation considerations. But his value judgements are thus fundamentally no more valid than the often more naive approach of the nature conservationist layman. Thus, the professional nature conservationist may, under certain circumstances, take comfort in the disappearance of a fen close to his home. This may not have been of very great value from a conservation point of view. It may have represented biotic communities that are still present at other sites

Fig. 40.1 Nature users. **(a)** Camping family at the edge of a nature reserve. **(b)** Elderly ladies in the nature reserve looking for a botanical photo motif. (Own recordings)

in the immediate vicinity. Or it may have been floristically degraded anyway. Only the resident of the residential environment can consider what this particular fen means to him and why he would not want to lose it. His or her perspective, possibly determined by proximity to home, recreational interests and integration into everyday life, is by no means subordinate to the nature conservation perspective; on the contrary, it complements it.

It would be nonsense to involve laypersons in a discussion about competing models in quantum physics or algorithmic approaches in chemical research. However, the expertise of professional nature conservation is of a different kind. Nature users (Fig. 40.1) must play a special evaluative role in it.

References

Adams JS, McShane TO (1996) The myth of wild Africa. University of California Press, Berkeley

Bayerische Akademie der Wissenschaften (1995) Bayerische Tropenforschung—Einst und jetzt. Rundgespräche der Kommission für Ökologie 10. Friedrich Pfeil, München

Behrens & Hoffmann (2013) Naturschutz-Geschichte(n). Lebenswege zwischen Ostseeküste und Erzgebirge. Steffen Verlag, Berlin

Bundesamt für Naturschutz (2012a) Daten zur Natur 2012. Bundesamt für Naturschutz, Bonn

Bundesamt für Naturschutz (2012b) Rote Karte für Isegrim. Nat Landsch 87(4):193

Bundesamt für Naturschutz (2012c) Hot Spot Wiese. Nat Landsch 87(6):277

Bundesamt für Naturschutz (2012d) Zahl des Monats. Nat Landsch 87(7):323

Bundesamt für Naturschutz (2013a) Bayerisches Löffelkraut und andere bedrohte Arten sichern. Nat Landsch 88(11):27

K.-D. Hupke, *Nature Conservation*,
https://doi.org/10.1007/978-3-662-66159-8

Bundesamt für Naturschutz (2014) Kampf gegen invasive Arten in der EU. Nat Landsch 89(2):94. (zitiert nach Niedersächsische Allgemeine vom 14.12.2013)

Bundesnaturschutzgesetz (2009) Deutscher Taschenbuch Verlag (ed) 2015. DTV, Munic

Czybulka et al (2018) Laubholz-Irrweg? Erwiderung auf: Wissenschaftlicher Beirat für Waldpolitik. Nat Landsch 93(7):344–345

Delius U (2016) Indigene Fischer fürchten Tiefseebergbau. Pogrom 297:46–49

Deutsche Bahn (2012) Artenwanderung—die stille Invasion. Mobil 3:72–80

Deutscher Taschenbuch Verlag (2015) Naturschutzrecht, München

Ellenberg H, Leuschner C (2010) Vegetation Mitteleuropas mit den Alpen. Eugen Ulmer, Stuttgart

Fiene C (2013) Wahrnehmung von Risiken aus dem globalen Klimawandel. Dissertation Abt. Geographie PH, Heidelberg

Geldmann J, Gonzáles-Varo JP (2018) Conserving honey bees does not help wildlife. High densities of managed honey bees can harm populations of wild pollinators. Science 359:392–393

Gigon A, von Rütte M (2014) Vorbehalte gegen den Naturschutz—Befragungsergebnisse und Hinweise zu Entgegnungen. Nat Landsch 89(7):310–316

Goren-Inbar N, Speth JD (eds) (2004) Human paleoecology in the Levantine corridor. Oxbow Books, Oxford

Gradmann R (1898) Das Pflanzenleben der Schwäbischen Alb mit Berücksichtigung der angrenzenden Gebiete Süddeutschlands, Bd 1. Verlag des Schwäbischen Albvereins, Tübingen

Grzimek B, Grzimek M (1959) Serengeti darf nicht sterben. Ullstein, Berlin - Frankfurt - Vienna

Grzimek B, Grzimek M (1961) Serengeti darf nicht sterben: 367.000 Tiere suchen einen Staat. Deutscher Bücherbund, Stuttgart

Hard G (1997) Spontane Vegetation und Naturschutz in der Stadt. Geogr Rundsch 49(10):562–568

Hejda M, de Bello F (2013) Impact of plant invasions on functional diversity in the vegetation of Central Europe. J Veg Sci 24:890–897

Hupke K-D (1981) Die Waldgrenze im Innerötztal (Tirol) Unpublished. Staatsexamensarbeit, U Stuttgart

Jax K (1994) Mosaik-Zyklus und Patch-dynamics: Synonyme oder verschiedene Konzepte? Eine Einladung zur Diskussion. Z Ökol Naturschutz 3:107–112

Kiemstedt H (1967) Zur Bewertung natürlicher Landschaftselemente für die Planung von Erholungsgebieten. Dissertation, Technische Hochschule Hannover

Kuckartz U, Grunenberg H (2002) Umweltbewusstsein in Deutschland 2002 Ergebnisse einer repräsentativen Bevölkerungsumfrage. Umweltbundesamt, Berlin

Kuckartz U, Rädiker S (2012) Das Bewusstsein über biologische Vielfalt in Deutschland: Wissen. Einstellung und Verhalten Nat Landsch 87(3):109–113

Kühne O (2013) Landschaftstheorie und Landschaftspraxis. Eine Einführung aus sozialkonstruktivistischer Perspektive. Springer, Wiesbaden

Lehmann J et al (2004) Amazonian dark earths. Origin—properties—management. Kluwer Academic Publishers, Alphen

Martin PS, Wright HE (eds) (1967) Pleistocene extinctions: the search for a cause. Yale University Press, New Haven

Minor Informationsbrief der "Arbeitsgemeinschaft Schildkröten" (2013) Straßenverkehr und Schildkröten. Minor 12(4):15

Möller A (1922) Der Dauerwaldgedanke Sein Sinn und seine Bedeutung. Springer, Berlin

Negi SS (2008) Biodiversity and its conservation in India. Status, threats and conservation, 2nd edn. Indus, New Delhi

Nentwig W (2010) Invasive Arten. Haupt, Bern

Reichholf J (2007) Stadtnatur Eine neue Heimat für Tiere und Pflanzen. Oekom, München

Reichholf J (2009) Die Zukunft der Arten. Neue ökologische Überraschungen. DTV, Frankfurt/M

Reisigl H, Pitschmann H (1958) Obere Grenzen von Flora und Vegetation in der Nivalstufe der zentralen Ötztaler Alpen (Tirol). Vegetatio 8:93–129

Runte A (2010) National parks. The American experience. Taylor Trade, Lanham

Schäfer A, Kowatsch A (2015) Gewässer und Auen—Nutzen für die Gesellschaft. Bundesamt f Naturschutz, Bonn

Scherzinger W (1996) Naturschutz im Wald. Qualitätsziele einer dynamischen Waldentwicklung. Eugen Ulmer, Stuttgart

Schlegelmilch K et al (2018) Naturbasierte Lösungen für Klimaschutz und Anpassung an den Klimawandel – Nutzen von Naturschutzmaßnahmen. Nat u Landsch 93:569–577

Schönborn W, Risse-Buhl U (2013) Lehrbuch der Limnologie. Schweizerbart, Stuttgart

Schönichen (1954) Naturschutz – Heimatschutz. Ihre Begründung durch Ernst Rudorff, Hugo Conwentz und ihre Vorläufer. Wissenschaftliche Verlagsgesellschaft, Stuttgart

Sebald O et al (eds) (1992) Die Farn- und Blütenpflanzen Baden-Württembergs, Bd 4. Eugen Ulmer, Stuttgart

Sebald O et al (1993) Die Farn- und Blütenpflanzen Baden-Württembergs. Bd. 1. Eugen Ulmer, Stuttgart

Sebald O et al (1996) Die Farn- und Blütenpflanzen Baden-Württembergs. Bd. 5. Eugen Ulmer, Stuttgart

Semmel A (1977) Grundzüge der Bodengeographie. Teubner, Stuttgart

Shackleton C, Shackleton S (2011) Non-timber forest products in the global context. Springer, Berlin

Shetler JB (2007) Imagining Serengeti. A history of landscape memory on Tanzania from earliest times to the present. Ohio University Press, Athens

Sieferle RP (1984) Fortschrittsfeinde? Opposition gegen Technik und Industrie von der Romantik bis zur Gegenwart. Beck, München

Steffan-Dewenter I, Tscharntke T (2000) Resource overlap and possible competition between honey bees and wild bees in Central Europe. Oecologia 122:288–296

Succow M, Jeschke L, Knapp HD (eds) (2012) Naturschutz in Deutschland. Ch Links, Berlin

Tesdorpf JC (1984) Landschaftsverbrauch. Begriffsbestimmung, Ursachenanalyse und Vorschläge zur Eindämmung. Dargestellt an Beispielen Baden-Württembergs. Verlag Dr. Tesdorpf, Berlin/Vilseck

Uhrmeister B, Reiff N, Falter R (1998) Rettet unsere Flüsse. Kritische Gedanken zur Wasserkraft. Pollner, Oberschleißheim b, München

von Carlowitz HC (1713) Sylvicultura oeconomica. J.F. Braun, Leipzig

Wilson JB (2014) Does the initial floristic composition model of secession really work? J Veg Sci 25(1):4

Wissenschaftlicher Beirat für Waldpolitik (ed) (2016) Klimaschutz in der Land- und Forstwirtschaft sowie den nachgelagerten Bereichen Ernährung und Holzverwendung. Bundesministerium für Ernährung und Landwirtschaft, Berlin

Wittig R, Niekisch M (2014) Biodiversität: Grundlagen, Gefährdung Schutz. Springer Spektrum, Berlin

Wrangham R (2009) Catching fire. How cooking made us human. Basic Books, New York

Index

K.-D. Hupke, *Nature Conservation*,
https://doi.org/10.1007/978-3-662-66159-8

B

O

P

S

T

GPSR Compliance

The European Union's (EU) General Product Safety Regulation (GPSR) is a set of rules that requires consumer products to be safe and our obligations to ensure this.

If you have any concerns about our products, you can contact us on ProductSafety@springernature.com

In case Publisher is established outside the EU, the EU authorized representative is:

Springer Nature Customer Service Center GmbH
Europaplatz 3
69115 Heidelberg, Germany

Batch number: 10732117

Printed by Printforce, the Netherlands